T0256718

Alan Turing's Systems of Logic

CLASS OF SERVICE DESIRED	
DOMESTIC	CABLE
TELEGRAM	FULL RATE
DAY LETTER	DEFERRED
NIGHT MESSAGE	CABLE LETTER
NIGHT LETTER	WEEK END LETTER

Patrons should check class of service desired; otherwise message will be transmitted as a full-rate communication.

WESTERN UNION

NEWCOMB CARLTON, PRESIDENT J. C. WILLEVER, FIRST VICE-PRESIDENT

Form 1206-A

NO.	CASH OR CHG.
CHECK	
TIME FILED	

Send the following message, subject to the terms on back hereof, which are hereby agreed to

September 10, 1936

TURING,
8 ENNISMORE AVENUE,
GUILDFORD, ENGLAND.

YOU ARE GRANTED ADMISSION TO PRINCETON GRADUATE SCHOOL

FOR COMING YEAR. LETTER FOLLOWS.

LUTHER EISENHART, DEAN

Note: A reply up to the amount of $1.50 is prepaid on this cable,
Send as cable night letter

24 July

King's College,
Cambridge.

Dear Mr Eisenhart,

The American Consulate in London tells me that I shall again need a letter in duplicate giving evidence that I am accepted as a student at Princeton, before I can be readmitted to the U.S. I should be grateful if you would send me such letters: it would be convenient if you would mention at the same time that I have been awarded a Fellowship which will cover all expenses; this would save my obtaining any other evidence about finances,

Yours sincerely

A. M. Turing

Alan Turing's Systems of Logic

THE PRINCETON THESIS

Edited and introduced by Andrew W. Appel

PRINCETON UNIVERSITY PRESS
PRINCETON AND OXFORD

Published by Princeton University Press,
41 William Street, Princeton, New Jersey 08540
In the United Kingdom: Princeton University Press, 6 Oxford Street, Woodstock, Oxfordshire OX20 1TW
press.princeton.edu

Second printing, and first paperback printing, 2014

Cloth ISBN 978-0-691-15574-6
Paper ISBN 978-0-691-16473-1
Library of Congress Control Number: 2012931772
British Library Cataloging-in-Publication Data is available

This book has been composed in Minion Pro
Printed on acid-free paper. ∞
Printed in the United States of America
10 9 8 7 6 5 4 3 2

Contents

Preface

Alan M. Turing, after his great result in 1936 discovering a universal model of computation and proving his incompleteness theorem, came to Princeton in 1936–38 and earned a PhD in mathematics. Before 1936 there were no universal computers. By 1955 there was not only a theory of computation, but there were real universal ("von Neumann") computers in Philadelphia, Cambridge (Massachusetts), Princeton, Cambridge (England), and Manchester. The new field of computer science had a remarkably short gestation.

The great engineers who built the first computers are well known: Konrad Zuse (Z3, Berlin, 1941); Tommy Flowers (Colossus, Bletchley Park, 1943); Howard Aiken (Mark I, Harvard, 1944); Prosper Eckert and John Mauchley (ENIAC, University of Pennsylvania, 1946).

But computer science is not just the construction of hardware. Who were the creators of the intellectual revolution underlying the theory of computers and computation?

Turing is very well known as a founder and pioneer of this discipline. In 1936 at the age of twenty-four he discovered the universal model of computation now known as the Turing machine; in 1938 he developed the notion of "oracle relativization"; in 1939–45 he was a principal figure in breaking the German Enigma ciphers using computational devices (though not "Turing machines"); in 1948 he invented the LU-decomposition method in numerical computation; in 1950 he foresaw the field of artificial intelligence and made remarkably accurate predictions about the future of computing and computers. And, of course, he famously committed suicide in 1954 after prosecution and persecution for practicing homosexuality in England.

But as significant as Turing is for the foundation of computer science, he was not the only scholar whose work in the 1930s led to the birth of this field.

In Fine Hall,[1] home in the 1930s of the Princeton Mathematics Department and the newly established Institute for Advanced Study, were mathematicians whose students would form a significant part of the new fields of computer science and operations research.

This volume presents the manuscript of Alan Turing's PhD thesis. It is accompanied by two introductory essays that explore both the work and the context of Turing's stay in Princeton. My essay elucidates the significance of Turing's work (and that of his adviser, Alonzo Church) for the field of computer science; Solomon Feferman's essay describes its significance for mathematics. Feferman also explains how to relate some of Turing's 1938 terminology to more current usage in the field. But on the whole, the notation and terminology in this field have been fairly stable: "Systems of Logic Based on Ordinals" is still readable as a mathematical and philosophical work.

Andrew W. Appel
Princeton, New Jersey

1 Fine Hall was built in 1930, named for the mathematician Henry Burchard Fine. During the 1930s it housed the Mathematics Department of Princeton University and the mathematicians (e.g., Gödel and von Neumann) and physicists (e.g., Einstein) of the Institute for Advanced Study. In 1939, the Institute moved to its own campus about a mile away from Princeton University's central campus. In 1969, the University's Mathematics Department moved to the new Fine Hall on the other side of Washington Road. The old building was renamed Jones Hall, in honor of its original donors, and now houses the departments of East Asian Studies and Near Eastern Studies.

OSWALD VEBLEN, chairman of the Princeton University Mathematics Department and first professor at the Institute for Advanced Study. His students include Alonzo Church (PhD 1927), and his PhD descendants through Philip Franklin (Princeton PhD 1921) via Alan Perlis (Turing Award 1966) include David Parnas, Zohar Manna, Kai Li, Jeannette Wing, and 500 other computer scientists. Veblen has more than 8000 PhD descendants overall. He helped oversee the development of the pioneering ENIAC digital computer in the 1940s.

(Photographer unknown, from the Shelby White and Leon Levy Archives Center, Institute for Advanced Study, Princeton, NJ, USA.)

ALONZO CHURCH, professor of mathematics, whose students include Alan Turing, Leon Henkin, Stephen Kleene, Martin Davis, Michael Rabin (Turing Award 1976), Dana Scott (Turing Award 1976), and Barkley Rosser, and whose PhD descendants include 3000 other mathematicians and computer scientists, among them Robert Constable, Edmund Clarke (Turing Award 2007), Allen Emerson (Turing Award 2007), and Les Valiant (Turing Award 2010).

(Photo from the Alonzo Church Papers. Department of Rare Books and Special Collection. Princeton University Library.)

SOLOMON LEFSHETZ, professor of mathematics, whose students include John McCarthy (Turing Award 1971), John Tukey, Ralph Gomory, and Richard Bellman (inventor of dynamic programming), and whose 6181 PhD descendants include John Nash (Nobel Prize 1994), Marvin Minsky (Turing Award 1969), Manuel Blum (Turing Award 1995), Barbara Liskov (Turing Award 2008), Gerald Sussman, Shafi Goldwasser, Umesh and Vijay Vazirani, Persi Diaconis, and Peter Buneman.

(Photo courtesy of the Princeton University Archives. Department of Rare Books and Special Collection. Princeton University Library.)

KURT GÖDEL, visitor to the Institute in 1933, 1934, and 1935, and professor at the Institute from 1940, had no students but had an enormous influence on the fields of mathematics and computer science. His 1931 incompleteness result—that it will never be possible to enumerate in logic the true statements of mathematics—stunned mathematicians and philosophers with its unexpectedness. His methods—the numerical encoding of syntax and the numerical processing of logic—set the stage for many techniques of computer science. Major results of Church, Kleene, Turing, and von Neumann clearly and explicitly owe much to Gödel.

(Photo from the Kurt Gödel Papers, the Shelby White and Leon Levy Archives Center, Institute for Advanced Study, Princeton, NJ, USA, on deposit at Princeton University.)

JOHN VON NEUMANN, at Princeton University from 1930 and professor at the Institute for Advanced Study from 1933, had only a few students (including the pioneer in parallel computer architecture Donald Gillies), but also had an enormous influence on the development of physics, mathematics, logic, economics, and computer science. In 1931 he was the first to recognize the significance of Gödel's work, and toward 1950 he brought Turing's ideas of program-as-data to the engineering of the first stored-program computers. Stored-program computers are called "von Neumann machines," and essentially all computers today are von Neumann machines.

(Photographer unknown, from the Shelby White and Leon Levy Archives Center, Institute for Advanced Study, Princeton, NJ, USA.)

The Birth of Computer Science at Princeton in the 1930s

ANDREW W. APPEL

The "Turing machine" is the standard model for a simple yet universal computing device, and Alan Turing's 1936 paper "On computable numbers . . . " (written while he was a fellow at Cambridge University) is the standard citation for the proof that some functions are not computable. But earlier in the same decade, Kurt Gödel at the Institute for Advanced Study in Princeton had developed the theory of recursive functions; Alonzo Church at Princeton University had developed the lambda-calculus as a model of computation; Church (1936) had just published his result that some functions are not expressible as recursive functions; and he had stated what we know as Church's Thesis: that the recursive functions characterize exactly the *effectively calculable* functions. In hindsight, the first demonstration that some functions are not computable was Church's.

It was only natural that the mathematician M. H. A. Newman (whose lectures on logic Turing had attended) should suggest that Turing come to Princeton to work with Church. Some of the greatest logicians in the world, thinking about the issues that in later decades became the foundation of computer science, were in Princeton's (old) Fine Hall in the 1930s: Gödel, Church, Stephen Kleene, Barkley Rosser, John von Neumann, and others. In fact, it is hard to imagine a more appropriate place for Turing to have pursued graduate study. After publishing his great result on computability, Turing spent two years (1938–38) at Princeton, writing his PhD thesis on "ordinal logics" with Church as his adviser.

If Turing was not the first to define a universal model of computable functions, why is the Turing machine the standard model? These three models—Gödel's recursive functions, Church's λ-calculus, and Turing's machine—were all proved equivalent in expressive power by Kleene (1936) and Turing (1937). But Turing's model is, most clearly of the three, a *machine*, with simple enough parts that one could imagine building it. As Solomon Feferman explains in his introduction to Turing's PhD thesis later in this volume, even Gödel was not convinced that either λ-calculus or his own model (recursive functions) was a sufficiently general representation of "computation" until he saw Turing's proof that unified recursive functions with Turing machines. That is, Church proved, and Turing independently re-proved, that some functions are not computable, but Turing's result was much more convincing about the *definition* of "computable."

Turing's "On computable numbers" convinced Gödel, and the rest of the world, in part because of the *philosophical* effort he put into that paper, as well as the mathematical effort. Turing described a process of computation as a human endeavor, or as a mechanical endeavor, in such a way that no matter which of these endeavors was dearest to the reader's heart, the result would come out the same: the Turing machine would express it. In contrast, it was not at all obvious that the Herbrand-Gödel recursive functions or the λ-calculus really constitutes the essence of "computation." We know that they do only because of the proof of equivalence with Turing machines.

The real computers of the 1940s and 1950s, like those of today, were never actually *Turing machines* with a finite control and an unbounded tape. But the electronic computers that *were* built, on both sides of the Atlantic, by von Neumann and others, were heavily (and explicitly) influenced by Turing's ideas, so that from the very beginning the field of computer science has often referred to computers in general as Turing machines—especially when considering their expressive power as universal computation devices.

What became of the other two models—recursive functions and λ-calculus? Most mathematicians working in computability theory use the theory of recursive functions; computer scientists working in computational complexity theory use both Turing machines and recursive functions. Turing himself used λ-calculus in his own PhD thesis, but, as Feferman explains,

> One reason that the reception of Turing's [PhD thesis] may have been so limited is that (no doubt at Church's behest) it was formulated in terms of the λ-calculus, which makes expressions for the ordinals and formal systems very hard to understand. He could instead have followed Kleene, who wrote in his retrospective history: "I myself, perhaps unduly influenced by rather chilly receptions from audiences around 1933–35 to disquisitions on λ-definability, chose, after general recursiveness had appeared, to put my work in that format. I cannot complain about my audiences after 1935."

For Feferman and Kleene, and for other mathematicians working in the field known as "recursive function theory," the particular *implementations* of functions (as described in λ-calculus) are rarely useful, and it is usually sufficient (and simpler) to talk more abstractly about the *existence* of implementations, that is, about definability and about enumerations of computable functions. Soare (1996) points out that the very name of the field (in mathematics) "recursive function theory" was invented by Kleene; Soare suggested "computability theory" as a more descriptive name for the field, and pointed out that Turing and Gödel used "computable" in preference to "recursive." Of course, Soare is both a mathematician and a computer scientist, and it is my impression that many of the latter used the term "computable" more frequently than "recursive" for decades before 1996, influenced (for example) by Martin Davis (PhD 1950 under Church).

So there were several models of computation, all known (by the end of the 1930s) to be equivalent: recursive functions, λ-calculus, Turing machines, and in fact others; for a few decades, mathematicians studied what can be represented as recursive functions, while the computer scientists studied what can be calculated by Turing machines.

But the λ-calculus did not disappear. In 1960 it became the explicit model of the Lisp programming language (invented by John McCarthy, 1927–2011; Turing Award 1971, PhD Princeton 1951 under Solomon Lefschetz). And λ-calculus is the implicit model of the Algol programming language (Perlis and Samelson 1958). Almost all programming languages in use today are descended from Lisp and Algol. Notions and mechanisms of variable binding, scope, functions, parameter passing, expressions, and type checking are all imported directly from Church's λ-calculus.

When the computers ("von Neumann machines") of the 1950s were built, with their (necessarily) sequential and mechanical universal control systems à la Turing, it was noticed that they were difficult to program. Programming these computers became easier with languages for specifying recursive functions (i.e., computations) that emphasized, to the degree possible with the technology of the time, *functions* (instead of procedures), *variables* (instead of registers), *binding* (instead of the "move" instruction), and *typed data* (instead of bit strings). All of this is from Church, and none of it is from Turing, Gödel, Kleene, or von Neumann.

Some mathematicians' criticisms of Church have to do with his reputation for pedanticism and excessive rigor: Hodges (1983, p. 119) writes that Turing "was reduced to attending Church's lectures, which he found ponderous and excessively precise." In part this reputation is undeserved. Feferman (1988, p. 120) writes that this "characterization of Church's style and personality" is "fair enough. . . . But it should be added that Church was (and is) noted for the great care and precision of his writing and lecturing, and these virtues probably benefited Turing—whose own writing was rough and ready and prone to minor errors." Robert Soare, who took classes from Church as an undergraduate at the beginning of the 1960s, says that Church's lectures on computability theory were indeed precise but "made the subject exciting"; Church was a better teacher than most math professors at Princeton.[1]

Still, Kleene and Feferman clearly agree that λ-calculus was not the most perspicuous vehicle for Turing to use in his PhD thesis, or for mathematicians to do many kinds of computability theory. This is because (typically) the mathematics they are doing depends only on the *computability* of a function,

1 Robert Soare, personal communication, December 12, 2011.

not on *which method* is used to compute it. In contrast, the engineers and programmers who have written programs from 1950 to the present are forced to write down a precise formulation of the function; otherwise we have *bugs*. So the notation for writing down representations of computable functions must be precise, and must also be both readable and writable (by humans and by other computer programs) both in the small and in the large. This is where the descendants of Church's notation work better than those of Turing's.

Some of the ways in which early programming languages differed from λ-calculus were forced by the limitations of early computers. Turing machines with an infinite tape and unbounded time can nicely simulate the λ-calculus. The slow computers of the 1960s and 1970s with their tiny memories forced programmers, even those who used Lisp and Algol, to split the difference between Church and von Neumann in how they wrote down their algorithms. But in the 1980s and 1990s, as computers became more powerful, it was possible to develop and apply programming languages (such as ML and Haskell[2]) that were even closer to Church's λ-calculus, and consciously so.

This brings me to the subject of Turing's Princeton PhD thesis, the content of the current volume. Here, Turing turns his attention from computation to logic. Gödel and Church would not have called themselves computer scientists: they were mathematical logicians; and even Turing, when he got his big 1936 result "On computable numbers," was answering a question in *logic* posed by Hilbert in 1928.

Turing's thesis, "Systems of Logic Based on Ordinals," takes Gödel's stunning incompleteness theorems as its point of departure. Gödel had shown that if a formal axiomatic system (of at least minimal expressive power) is consistent, then it cannot be complete. And not only is the system incomplete, but the formal statement of the consistency of the system is true and unprovable if the system is consistent. Thus if we already have informal or intuitive reasons for accepting the axioms of the system as true, then we ought to accept the statement of its consistency as a new axiom. And then we can apply the same considerations to the new system; that is, we can iterate the process of adding consistency statements as new axioms. In his thesis, Turing investigated that process systematically by iterating it into the constructive transfinite, taking

2 Named after another great logician, Haskell Curry, who was also visiting Princeton in 1938.

unions of logical systems at limit ordinal notations. His main result was that one can thereby overcome incompleteness for an important class of arithmetical statements (though not for all).

It is clear that Turing regards the formalization of mathematics as a desirable goal. He excuses himself at one point (on pp. 9–10 of the manuscript):

> There is another point to be made clear in connection with the point of view we are adopting. It is intended that all proofs that are given [in this thesis] should be regarded no more critically than proofs in classical analysis. The subject matter, roughly speaking, is constructive systems of logic, but as the purpose is directed towards choosing a particular constructive system of logic for practical use; an attempt at this stage to put our theorems into constructive form would be putting the cart before the horse.

Here it is clear that Turing is a logician and not just a great mathematician; few mathematicians believe that it would be a useful purpose to choose a constructive system of logic for *practical* use, and no ordinary mathematician would excuse himself for being no more rigorous than a mathematician.

Just as one of the strengths of Turing's great 1936 paper was its philosophical component—in which he explains the motivation for the Turing machine as a model of computation—here in the PhD thesis he is also motivated by philosophical concerns, as in section 9 (p. 60 of the manuscript):

> We might hope to obtain some intellectually satisfying system of logical inference (for the proof of number theoretic theorems) with some ordinal logic. Gödel's theorem shows that such a system cannot be wholly mechanical, but with a complete ordinal logic we should be able to confine the non-mechanical steps entirely to verifications that particular formulae are ordinal formulae.

Turing greatly expands on these philosophical motivations in section 11 of the thesis. His program, then, is this: We wish to reason in some logic, so that our proofs can be mechanically checked (for example, by a Turing machine). Thus we don't need to trust our students and journal-referees to check our proofs. But no (sufficiently expressive) logic can be complete, as Gödel showed. If we are using a given logic, sometimes we may want to reason about statements unprovable in that logic. Turing's proposal is to use an ordinal logic sufficiently high in the hierarchy; checking proofs in that logic will be completely mechanical, but the one "intuitive" step remains of verifying ordinal formulas.

Unfortunately, it is not at all clear that verifying ordinal formulas is in any way "intuitive." Feferman (1988, sec. 6) estimates that "the demand on 'intuition' in recognizing 'which formulae are ordinal formulae' is somewhat greater than Turing suggests." Feferman concludes his essay included in this volume with a mention of his and Kreisel's subsequent approaches to this problem, between 1958 and 1970.

Turing, in the thesis, recognizes significant problems with his ordinal logics, which can be summarized by his statement (manuscript, p. 73) that "with almost any reasonable notation for ordinals, completeness is incompatible with invariance" (and see also Feferman's essay).

But the PhD thesis contains, almost as an aside, an enormously influential mathematical insight. Turing invented the notion of oracles, in which one kind of computer consults from time to time, in an explicitly axiomatized way, a more powerful kind. Oracle computations are now an important part of the tool kit of both mathematicians and computer scientists working in computability theory and computational complexity theory (see Feferman 1992; Soare 2009). This method may actually be the most significant result in Turing's PhD thesis.

So the thesis exhibits Turing as logician. Alonzo Church also continued to be a logician, as in 1940 he published "A Formulation of the Simple Theory of Types," setting out the system now known as higher-order logic. As a practical means of actually doing mechanized reasoning, Turing's 1938 result was not nearly as influential as Church's higher-order logic.

In many other fields of engineering, such as the construction of bridges, chemical processes, or photonic circuits, the applicable mathematics is from analysis or quantum mechanics (see Wigner 1960, "The Unreasonable Effectiveness of Mathematics in the Natural Sciences"). But software does not (principally) rely on continuous or quantum artifacts of the natural world, where that kind of math works so well. Instead, software follows the discrete logic of bits, and it obeys axioms specified by the engineers who designed the instruction-set architecture of the computer, and by those who specified the semantics of the programming languages. Thus the applicable mathematics is, in fact, logic (see Halpern et al. 2001, "On the Unusual Effectiveness of Logic in Computer Science").

It might seem that the Boolean algebra of bits is simpler than real analysis, but the problem is that software systems are so complex that the reasoning is difficult. Thus in the twenty-first century many computer scientists do mechanized formal reasoning, and the most significant application domain for mechanized proof is in the verification of computer software itself. Software is large and complex, and for at least some software it is very desirable that it conform to a given formal specification. The theorems and proofs are too large for us to reliably build and maintain by hand, so we mechanize.

Mechanized proof comes in two flavors; the first flavor is fully automated. *Automated theorem proving* is the use of computer programs to find proofs automatically. *Automatic static analysis* is the use of computer programs to calculate behavioral properties of other computer programs, sometimes by calling upon automated theorem provers as subroutines to decide the validity of logical propositions. Do not be frightened by Turing's result that this problem is uncomputable; his result is simply that no automated procedure can decide the provability of *every* mathematical proposition, and no automated procedure can test nontrivial properties of *every* other program.[3] We do not need to prove *every* theorem or analyze *every* program; it will suffice to automatically prove many useful theorems, or analyze useful programs. Some automated provers work in undecidable logics, and (therefore) sometimes fail to find the proof. In those cases, the user is expected to simplify or reformulate the theorem as necessary, or provide hints. We would not ask Fermat to reformulate his Last Theorem for the convenience of Wiles; but when the theorem is "This horrible program meets its specification," we might well rewrite the program to make it more easily reasoned about. Other automated provers work in decidable logics—for example, Presburger arithmetic or Boolean satisfiability.

3 Actually, this generalization of Turing's 1936 result about halting is known as Rice's theorem (1953).

Do not be frightened by Cook's result (1971) that satisfiability is NP-complete; that result is simply that no (known) automated procedure can solve *every* instance in polynomial time. In practice, SAT-solvers are now a big industry; they are quite effective in solving the actual cases that come up in theorem-proving applications. (Of course, SAT-solvers are not so effective in solving problems that arise from *deliberately* intractable problems, such as cryptography.) The extension of SAT-solvers to SMT (satisfiability modulo theories) is also now a big academic and commercial industry. Many of these solvers use variants of the Davis-Putnam algorithm for resolution theorem proving, discovered in 1960 by Martin Davis (PhD 1950 under Church) and Hilary Putnam (PhD UCLA 1951; in 1960 a colleague of Church's at Princeton).

The other flavor of mechanized proof is the use of computer programs to *check* proofs automatically, and to *assist* in the bureaucratic details of their construction. These are the proof assistants. One of the earliest of these was Robin Milner's LCF (Logic for Computable Functions) system (Gordon et al. 1979). Milner was influenced by the work of Church and by that of Dana Scott (PhD 1958 under Church), Christopher Strachey (a fellow student of Turing's at Cambridge, and one of the first to program the ACE computer in 1951), and Peter Landin (a student of Strachey's). Strachey, Landin, and Milner, all British computer scientists, were important figures in the application of Church's λ-calculus and logic to the design of programming languages and formal methods for reasoning about them.

Although some proof assistants use first-order logics (i.e., logics where each quantifier ranges over elements of a particular fixed type), for the expression of mathematical ideas it is much more convenient to use higher-order logics (i.e., where the type of a quantifier can itself be a variable bound in an outer scope). One of the earliest higher-order logics is Church's "simple theory of types" (1940), but even more expressive (and, to my taste, more useful) logics have dependent types, where the type of one variable may depend on the value of another. Such logics include LF (the Logical Framework) and CoC (the Calculus of Constructions). Proof assistants such as HOL (using the simple theory of types), Twelf (using LF), and Coq (using CoC) are now routinely used to specify and prove substantial theorems about computers and computer programs.

Not only theorems about software; sometimes these proof assistants are even used to prove theorems in mathematics. Georges Gonthier (2008) used Coq to implement a proof of the four-color theorem end-to-end in "Church/Turing-style" fully formal logic. Gonthier's implementation improved on the 1976 proof by Kenneth Appel and Wolfgang Haken that relied in part on "von Neumann–style" Fortran programs to calculate reducibility and in part on "Pythagoras-style" traditional mathematics. (In 1976 the reaction of some mathematicians was to distrust those parts of Appel and Haken's proof that were calculated by computer, whereas the reaction of some computer scientists was to distrust the parts that were checked only "by hand.") In the twenty-first century, computer programs that prove mathematical theorems are expected themselves to be formalized within a mechanically checked logical system. Thomas Hales (2005) proved the Kepler conjecture about sphere packing, using computer programs written in Mathematica and C++, about which the referees were "99% certain." In order to reach 100%, Hales's current project (nearly complete) is to formalize this proof in the HOL Light proof assistant.

In Cambridge, Turing (1936) had brilliant, unprecedented ideas about the nature of computation. He was certainly not the first to build an actual computer; there was already work in progress at (for example) the University of Iowa. But when Turing came to Princeton to work with Church, in the orbit of Gödel, Kleene, and von Neumann,[4] among them they founded a field of computer science that is firmly grounded in logic. In some of Turing's other work (1950) he foresees the field (now within computer science) of artificial intelligence. But in his PhD thesis he makes it clear that he looks to a day when, in proving mathematical theorems, "the strain put on the intuition should be a minimum" (manuscript, page 83). That is, to the extent possible, every step in a proof should be mechanically checkable. We all know the Church-Turing thesis: that no realizable computer will be able to compute more functions than λ-calculus or a Turing machine. But in reading Turing's "Systems of Logic ..." we can see quite clearly another kind of Church-Turing thesis, that came

4 Gödel was away from Princeton during Turing's time here, and Kleene had already finished his PhD and left; but clearly they had an enormous influence on Turing's PhD thesis. Turing worked with von Neumann during his time at Princeton, but on other kinds of mathematics than logic and computation.

half a century later as a consequence of their work: mathematical reasoning *can* be done, and often *should* be done, in mechanizable formal logic.

BIBLIOGRAPHY

Church, Alonzo (1936), An unsolvable problem of elementary number theory, *American Journal of Mathematics* vol. 58, pp. 345–363.

Church, Alonzo (1940), A formulation of the simple theory of types, *Journal of Symbolic Logic* vol. 5, pp. 56–68.

Cook, Stephen A. (1971), The complexity of theorem-proving procedures, in *Proceedings 3rd ACM Symposium on the Theory of Computing*, pp. 151–158.

Davis, Martin, and Hilary Putnam (1960), A computing procedure for quantification theory, *Journal of the ACM* vol. 7, number 3, pp. 201–215.

Feferman, Solomon (1988), Turing in the land of O(z), in R. Herken, ed., *The Universal Turing Machine: a Half-Century Survey*, Oxford University Press.

Feferman, Solomon (1992), Turing's "oracle": From absolute to relative computability—and back, in J. Echeverria, et al., eds., *The Space of Mathematics,* Walter de Gruyter, pp. 314–348 .

Gonthier, Georges (2008), Formal proof—the four-color theorem, *Notices of the AMS* vol. 55, number 11, pp. 1382–1393.

Gordon, Michael J., Robin Milner, and Christopher P. Wadsworth (1979), *Edinburgh LCF: a Mechanised Logic of Computation,* Springer-Verlag.

Hales, Thomas C. (2005), A proof of the Kepler conjecture, *Annals of Mathematics* vol. 162, pp. 1065–1185.

Halpern, Joseph Y., Robert Harper, Neil Immerman, Phokion G. Kolaitis, Moshe Y. Vardi, and Victor Vianu (2001), On the unusual effectiveness of logic in computer science, *Bulletin of Symbolic Logic* vol. 7, number 2, pp. 213–236.

Hodges, Andrew (1983), *Alan Turing: the Enigma,* Burnett Books. (New American edition, Princeton University Press 2012.)

Kleene, S. C. (1936), λ-definability and recursiveness, *Duke Mathematical Journal* vol. 2, pp. 340–353.

McCarthy, John (1960), Recursive functions of symbolic expressions and their computation by machine, Part I, *Communications of the ACM* vol. 3, number 4, pp. 184–195.

Perlis, A. J., and K. Samelson (1958), Preliminary report: international algebraic language, *Communications of the ACM* vol. 1, number 12, pp. 8–22.

Rice, H. G. (1953), Classes of recursively enumerable sets and their decision problems, *Transactions of the American Mathematical Society* vol. 74, pp. 358–366.

Soare, Robert I. (1996), Computability and recursion, *Bulletin of Symbolic Logic* vol. 2, pp. 284–321.

Soare, Robert I. (2009), Turing oracle machines, online computing, and three displacements in computability theory, *Annals of Pure and Applied Logic*, vol. 160, issue 3, pp. 368–399.

Turing, A. M. (1936), On computable numbers, with an application to the Entscheidungsproblem, *Proceedings of the London Mathematical Society*, ser. 2, vol. 42, pp. 230–265.

Turing, A. M. (1937), Computability and λ-definability, *Journal of Symbolic Logic* vol. 2, number 4, pp. 153–163.

Turing. A. M. (1950), Computing machinery and intelligence, *Mind* vol. 59, pp. 433–460.

Wigner, Eugene P. (1960), The unreasonable effectiveness of mathematics in the natural sciences, *Communications on Pure and Applied Mathematics* vol. 13, number 1, pp. 1–14.

Turing's Thesis

SOLOMON FEFERMAN

In the sole extended break from his life and varied career in England, Alan Turing spent the years 1936–1938 doing graduate work at Princeton University under the direction of Alonzo Church, the doyen of American logicians. Those two years sufficed for him to complete a thesis and obtain the Ph.D. The results of the thesis were published in 1939 under the title "Systems of logic based on ordinals" [23]. That was the first systematic attempt to deal with the natural idea of overcoming the Gödelian incompleteness of formal systems by iterating the adjunction of statements—such as the consistency of the system—that "ought to" have been accepted but were not derivable; in fact these kinds of iterations can be extended into the transfinite. As Turing put it beautifully in his introduction to [23]:

> The well-known theorem of Gödel (1931) shows that every system of logic is in a certain sense incomplete, but at the same time it indicates means whereby from a system L of logic a more complete system L' may be obtained. By repeating the process we get a sequence L, $L_1 = L'$, $L_2 = L'_1$ … each more complete than the preceding. A logic L_ω may then be constructed in which the provable theorems are the totality of theorems provable with the help of the logics L, L_1 , L_2 , … Proceeding in this way we can associate a system of logic with any constructive ordinal. It may

> be asked whether such a sequence of logics of this kind is complete in the sense that to any problem A there corresponds an ordinal α such that A is solvable by means of the logic L_α.

Using an ingenious argument in pursuit of this aim, Turing obtained a striking yet equivocal partial completeness result that clearly called for further investigation. But he did not continue that himself, and it would be some twenty years before the line of research he inaugurated would be renewed by others. The paper itself received little attention in the interim, though it contained a number of original and stimulating ideas and though Turing's name had by then been well established through his earlier work on the concept of effective computability.

Here, in brief, is the story of what led Turing to Church, what was in his thesis, and what came after, both for him and for the subject.[1]

FROM CAMBRIDGE TO PRINCETON

As an undergraduate at King's College, Cambridge, from 1931 to 1934, Turing was attracted to many parts of mathematics, including mathematical logic. In 1935 Turing was elected a fellow of King's College on the basis of a dissertation in probability theory, *On the Gaussian error function*, which contained his independent rediscovery of the central limit theorem. Earlier in that year he began to focus on problems in logic through his attendance in a course on that subject by the topologist M. H. A. (Max) Newman. One of the problems from Newman's course that captured Turing's attention was the *Entscheidungsproblem*, the question whether there exists an effective method to decide, given any well-formed formula of the pure first-order predicate calculus, whether or not it is valid in all possible interpretations (equivalently, whether or not its negation is satisfiable in some interpretation). This had been solved in the affirmative for certain special classes of formulas, but the general problem was

1 I have written about this at somewhat greater length in [10]; that material has also been incorporated as an introductory note to Turing's 1939 paper in the volume, *Mathematical Logic* [25] of his collected works. In its biographical part I drew to a considerable extent on Andrew Hodges' superb biography, *Alan Turing: The Enigma* [16].

still open when Turing began grappling with it. He became convinced that the answer must be negative, but that in order to demonstrate the impossibility of a decision procedure, he would have to give an exact mathematical explanation of what it means to be computable by a strictly mechanical process. He arrived at such an analysis by mid-April 1936 via the idea of what has come to be called a *Turing machine*, namely an idealized computational device following a finite table of instructions (in essence, a program) in discrete effective steps without limitation on time or space that might be needed for a computation. Furthermore, he showed that even with such unlimited capacities, the answer to the general *Entscheidungsproblem* must be negative. Turing quickly prepared a draft of his work entitled "On computable numbers, with an application to the *Entscheidungsproblem*"; Newman was at first skeptical of Turing's analysis but then became convinced and encouraged its publication.

Neither Newman nor Turing were aware at that point that there were already two other proposals under serious consideration for analyzing the general concept of effective computability: one by Gödel called *general recursiveness*, building on an idea of Herbrand, and the other by Church, in terms of what he called the *λ-calculus*.[2] In answer to the question, "Which functions of natural numbers are effectively computable?", the Herbrand-Gödel approach was formulated in terms of finite systems of equations from which the values of the functions are to be deduced using some elementary rules of inference; since the functions to be defined can occur on both sides of the equations, this constitutes a general form of recursion. Gödel explained this in lectures on the incompleteness results during his visit to the Princeton Institute for Advanced Study in 1934, lectures that were attended by Church and his students Stephen C. Kleene and J. Barkley Rosser. But Gödel regarded general recursiveness only as a "heuristic principle" and was not himself willing to commit to that proposed analysis. Meanwhile Church had been exploring a different answer to the same question in terms of his λ-calculus—a fragment of a quite general formalism for the foundation of mathematics, whose fundamental notion

2 The development of ideas about computability in this period by Herbrand, Gödel, Church, Turing, and Post has been much written about and can only be touched on here. For more detail I recommend the article by Kleene [17] and the articles by Hodges, Kleene, Gandy, and Davis in Part I of Herken's collection [15], among others. One of the many good online sources with further links is at http://plato.stanford.edu/entries/church-turing/, by B. J. Copeland.

is that of arbitrary functions rather than arbitrary sets. The "λ" comes from Church's formalism according to which if $t[x]$ is an expression with one or more occurrences of a variable x, then $\lambda x.t[x]$ is supposed to denote a function f whose value $f(s)$ for each s is the result, $t[s/x]$, of substituting s for x throughout t.[3] In the λ-calculus, function application of one expression t to another s as argument is written in the form ts. Combining these, we have the basic evaluation axiom: $(\lambda x.t[x])s = t[s/x]$.

Using a representation of the natural numbers in the λ-calculus, a function f is said to be λ-*definable* if there is an expression t such that for each pair of numerals $\boldsymbol{n}$ and $\boldsymbol{m}$, $\boldsymbol{tn}$ evaluates out to $\boldsymbol{m}$ if and only if $f(n) = m$. In conversations with Gödel, Church proposed λ-definability as the precise explanation of effective computability ("Church's Thesis"), but in Gödel's view that was "thoroughly unsatisfactory". It was only through a chain of equivalences that ended up with Turing's analysis that Gödel later came to accept it, albeit indirectly. The first link in the chain was forged with the proof by Church and Kleene that λ-definability is equivalent to general recursiveness. Thus when Church finally announced his "Thesis" in published form in 1936 [1], it was in terms of the latter. In that paper, Church applied his thesis to demonstrate the effective unsolvability of various mathematical and logical problems, including the decision problem for sufficiently strong formal systems. And then in his follow-up paper [2] submitted April 15, 1936—just around the time Turing was showing Newman *his* draft—Church proved the unsolvability of the *Entscheidungsproblem* for logic. When news of this work reached Cambridge a month later, the initial reaction was great disappointment at being scooped, but it was agreed that Turing's analysis was sufficiently different to still warrant publication. After submitting it for publication toward the end of May 1936, Turing tacked on an appendix in August of that year in which he sketched the proof of equivalence of computability by a machine in his sense with that of λ-definability, thus forging the second link in the chain of equivalences [21].

In Church's 1937 review of Turing's paper, he wrote:

> As a matter of fact, there is involved here the equivalence of three differ-

3 One must avoid the "collision" of free and bound variables in the process, i.e., no free variable z of s must end up within the scope of a "λz"; this can be done by renaming bound variables as necessary.

> ent notions: computability by a Turing machine, general recursiveness in the sense of Herbrand-Gödel-Kleene, and λ-definability in the sense of Kleene and the present reviewer. Of these, the first has the advantage of making the identification with effectiveness in the ordinary (not explicitly defined) sense evident immediately. . . . The second and third have the advantage of suitability for embodiment in a system of symbolic logic.[4]

Thus was born what is now called the *Church-Turing Thesis*, according to which the effectively computable functions are exactly those computable by a Turing machine.[5] The (Church-)Turing Thesis is of course not to be confused with Turing's thesis under Church, our main subject here.

TURING IN PRINCETON

On Newman's recommendation, Turing decided to spend a year studying with Church, and he applied for one of Princeton's Procter fellowships. In the event, he did not succeed in obtaining it, but even so he thought he could manage on his fellowship funds from King's College of 300 pounds per annum, and so Turing came to Princeton at the end of September 1936. The Princeton mathematics department had already been a leader on the American scene when it was greatly enriched in the early 1930s by the establishment of the Institute for Advanced Study. The two shared Fine Hall until 1940, so that the lines between them were blurred and there was significant interaction. Among the mathematical leading lights that Turing found on his arrival were Einstein, von Neumann, and Weyl at the Institute and Lefschetz in the department; the visitors that year included Courant and Hardy. In logic, he had hoped to find—besides Church—Gödel, Bernays, Kleene, and Rosser. Gödel had indeed commenced a second visit in the fall of 1935 but left after a brief period due to illness; he was not to return until 1939. Bernays (noted as Hilbert's collaborator on his consistency program) had visited 1935–36, but did not visit the States again until after the war. Kleene and Rosser had received their

4 Church's review appeared in *J. Symbolic Logic* 2 (1937), 42–43.

5 Gödel accepted the Church-Turing Thesis in that form in a number of lectures and publications thereafter.

Ph.D.s by the time Turing arrived and had left to take positions elsewhere. So he was reduced to attending Church's lectures, which he found ponderous and excessively precise; by contrast, Turing's native style was rough-and-ready and prone to minor errors, and it is a question whether Church's example was of any benefit in this respect. They met from time to time, but apparently there were no sparks, since Church was retiring by nature and Turing was somewhat of a loner.

In the spring of 1937, Turing worked up for publication a proof in greater detail of the equivalence of machine computability with λ-definability [22]. He also published two papers on group theory, including one on finite approximations of continuous groups that was of interest to von Neumann (cf. [24]). Luther P. Eisenhart, who was then head of the mathematics department, urged Turing to stay on for a second year and apply again for the Procter fellowship (worth US$2,000 p.a.). This time, supported by von Neumann who praised his work on almost periodic functions and continuous groups, Turing succeeded in obtaining the fellowship, and so decided to stay the extra year and do a Ph.D. under Church. Proposed as a thesis topic was the idea of ordinal logics that had been broached in Church's course as a way to "escape" Gödel's incompleteness theorems.

Turing, who had just turned 25, returned to England for the summer of 1937, where he devoted himself to three projects: finishing the computability/λ-definability paper, ordinal logics, and the Skewes number. As to the latter, Littlewood had shown that $\pi(x) - \mathrm{li}(x)$ changes sign infinitely often, with an argument by cases, according to whether the Riemann Hypothesis is true or not; prior to that it had been conjectured that $\pi(n) < \mathrm{li}(n)$ for all n, in view of the massive numerical evidence into the billions in support of that.[6] In 1933 Skewes had shown that $\mathrm{li}(n) < \pi(n)$ for some $n < 10_3(34)$ (triple exponential to the base 10) if the Riemann Hypothesis is true. Turing hoped to lower Skewes' bound or eliminate the Riemann Hypothesis; in the end he thought he had succeeded in doing both and prepared a draft but did not publish his work.[7] He was to have a recurring interest in the R.H. in the following years,

6 $li(x)$ is the (improper) integral from 0 to x of $1/\log x$ and is asymptotic to $\pi(x)$, the number of primes $< x$.

7 A paper based on Turing's ideas, with certain corrections, was published after his death by Cohen and Mayhew [4].

including devising a method for the practical computation of the zeros of the Riemann zeta function as explained in the article by Andrew R. Booker in this issue of the *Notices*. Turing also made good progress on his thesis topic and devoted himself full time to it when he returned to Princeton in the fall, so that he ended up with a draft containing the main results by Christmas of 1937. But then he wrote Philip Hall in March 1938 that the work on his thesis was "proving rather intractable, and I am always rewriting part of it."[8] Later he wrote that "Church made a number of suggestions which resulted in the thesis being expanded to an appalling length." One can well appreciate that Church would not knowingly tolerate imprecise formulations or proofs, let alone errors, and the published version shows that Turing went far to meet such demands while retaining his distinctive voice and original ways of thinking. Following an oral exam in May, on which his performance was noted as "Excellent", the Ph.D. itself was granted in June 1938. Turing made little use of the doctoral title in the following years, since it made no difference for his position at Cambridge. But it could have been useful for the start of an academic career in America. Von Neumann thought sufficiently highly of his mathematical talents to offer Turing a position as his assistant at the Institute. Curiously, at that time von Neumann showed no knowledge or appreciation of his work in logic. It was not until 1939 that he was to recognize the fundamental importance of Turing's work on computability. Then, toward the end of World War II, when von Neumann was engaged in the practical design and development of general purpose electronic digital computers in collaboration with the ENIAC team, he was to incorporate the key idea of Turing's universal computing machine in a direct way.[9]

Von Neumann's offer was quite attractive, but Turing's stay in Princeton had not been a personally happy one, and he decided to return home despite the uncertain prospects aside from his fellowship at King's and in face of the brewing rumors of war. After publishing the thesis work he did no more on that topic and went on to other things. Not long after his return to England, he joined a course at the Government Code and Cypher School, and that was

8 Hodges [16], p. 144.

9 Its suggested implementation is in the *Draft report on the EDVAC* put out by the ENIAC team and signed by von Neumann; cf. Hodges [16], pp. 302–303; cf. also ibid., p. 145, for von Neumann's appreciation by 1939 of the significance of Turing's work.

to lead to his top secret work during the war at Bletchley Park on breaking the German Enigma Code. This fascinating part of the story is told in full in Hodges' biography [16], as is his subsequent career working to build actual computers, promote artificial intelligence, theorize about morphogenesis, and continue his work in mathematics. Tragically, this ended with his death in 1954, a probable suicide.

THE THESIS: ORDINAL LOGICS[10]

What Turing calls a *logic* is nowadays more usually called a *formal system*, i.e., one prescribed by an effective specification of a language, set of axioms and rules of inference. Where Turing used "L" for logics I shall use "S" for formal systems. Given an effective description of a sequence $\langle S_n \rangle_{n \in N}$ ($N = \{0, 1, 2, \ldots\}$) of formal systems all of which share the same language and rules of inference, one can form a new system $S_\omega = \cup\, S_n$ $(n \in N)$, by taking the effective union of their axiom sets. If the sequence of S_n's is obtained by iterating an effective passage from one system to the next, then that iteration can be continued to form $S_{\omega+1}$, ... and so on into the transfinite. This leads to the idea of an effective association of formal systems S_α with ordinals α. Clearly that can be done only for denumerable ordinals, but to deal with limits in an effective way, it turns out that we must work not with ordinals per se, but with *notations for ordinals*. In 1936, Church and Kleene [3] had introduced a system O of constructive ordinal notations, given by certain expressions in the λ-calculus. A variant of this uses numerical codes a for such expressions and associates with each $\alpha \in O$ a countable ordinal $|\alpha|$. For baroque reasons, 1 was taken as the notation for 0, 2^a as a notation for the successor of $|a|$, and $3 \bullet 5^e$ for the limit of the sequence $|a_n|$, when this sequence is strictly increasing and when e is a code of a computable function $\hat{e}$ with $\hat{e}(n) = a_n$ for each $n \in N$. The least ordinal not of the form $|a|$ for some $a \in O$ is the analogue, in terms of effective computability, of the least uncountable ordinal ω_1 and is usually denoted by ω_1^{CK}, where "CK" refers to Church and Kleene. By an *ordinal logic* $S^* = \langle S_a \rangle_{a \in O}$ is

10 The background to the material of this section in Gödel's incompleteness theorems is explained in my piece for the *Notices* [11].

meant any means of effectively associating with each $a \in O$ a formal system S_a. Note, for example, that there are many ways of forming a sequence of notations a_n whose limit is ω, given by all the different effectively computable strictly increasing subsequences of N. So at limit ordinals $\alpha < \omega_1^{CK}$ we will have infinitely many representations of α and thus also for its successors. An ordinal logic is said to be *invariant* if whenever $|a| = |b|$ then S_a and S_b prove the same theorems.

In general, given any effective means of passing from a system S to an extension S' of S, one can form an ordinal logic $S^* = \langle S_a \rangle_{a \in O}$ which is such that for each $a \in O$ and $b = 2^a$ the successor of a, $S_b = S'_a$, and is further such that whenever $a = 3 \cdot 5^e$ then S_a is the union of the sequence of $S_{\hat{e}(n)}$ for each $n \in N$. In particular, for systems whose language contains that of Peano Arithmetic PA, one can take S' to be $S \cup \{Con_S\}$, where Con_S formalizes the consistency statement for S; the associated ordinal logic S^* thus iterates adjunction of consistency through all the constructive ordinal notations. If one starts with PA as the initial system it may be seen that each S_a is consistent and so S'_a is strictly stronger than S_a by Gödel's second incompleteness theorem. The consistency statements are expressible in $\forall$("for all")-form, i.e., $\forall x R(x)$ where R is an effectively decidable predicate. So a natural question to raise is whether S^* is complete for statements of that form, i.e., whether whenever $\forall x R(x)$ is true in N then it is provable in S_a for some $a \in O$. Turing's main result for this ordinal logic was that that is indeed the case, in fact one can always choose such an a with $|a| = \omega + 1$. His ingenious method of proof was, given R, to construct a sequence $\hat{e}(n)$ that denotes n as long as $(\forall x \leq n)R(x)$ holds and that jumps to the successor of $3 \cdot 5^e$ when $(\exists x \leq n)\neg R(x)$.[11] Let $b = 3 \cdot 5^e$ and $a = 2^b$. Now if $\forall x R(x)$ is true, $b \in O$ with $|b| = \omega$. In S_a we can reason as follows: if $\forall x R(x)$ were not true then S_b would be the union of systems that are eventually the same as S_a, so S_b would prove its own consistency and hence, by Gödel's theorem, would be inconsistent. But S_a proves the consistency of S_b, so we must conclude that $\forall x R(x)$ holds after all.

Turing recognized that this completeness proof is disappointing because it shifts the question of whether a $\forall$-statement is true to the question whether a

11 Note that e is defined in terms of itself; this is made possible by Kleene's index form of the recursion theorem.

number a actually belongs to O. In fact, the general question, given a, is $a \in O$?, turns out to be of higher logical complexity than any arithmetical statement, i.e., one formed by the unlimited iteration of universal and existential quantifiers, $\forall$ and $\exists$. Another main result of Turing's thesis is that for quite general ordinal logics, S^* can't be both complete for statements in $\forall$-form and invariant. It is for these reasons that above I called his completeness results equivocal. Even so, what Turing really hoped to obtain was completeness for statements in $\forall\exists$ ("for all, there exists")-form. His reason for concentrating on these, which he called "number-theoretical problems", rather than considering arithmetical statements in general, is not clear. This class certainly includes many number-theoretical statements (in the usual sense of the word) of mathematical interest, e.g., those, such as the twin prime conjecture, that say that an effectively decidable set C of natural numbers is infinite. Also, as Turing pointed out, the question whether a given program for one of his machines computes a total function is in $\forall\exists$-form. Of special note is his proof ([23], sec. 3) that the Riemann Hypothesis is a number-theoretical problem in Turing's sense. This was certainly a novel observation for the time; actually, as shown by Georg Kreisel years later, it can even be expressed in $\forall$-form.[12] On the other hand, Turing's class of number-theoretical problems does not include such statements as finiteness of the number of solutions of a diophantine equation ($\exists\forall$) or the statement of Waring's problem ($\forall\exists\forall$).

In section 4 Turing introduced a new idea that was to change the face of the general theory of computation (also known as recursion theory) but the only use he made of it there was curiously inessential. His aim was to produce an arithmetical problem that is not number-theoretical in his sense, i.e., not in $\forall\exists$-form. This is trivial by a diagonalization argument, since there are only countably many effective relations $R(x, y)$ of which we could say that $\forall x \exists y R(x, y)$ holds. Turing's way of dealing with this, instead, is through the new notion of computation relative to an *oracle*. As he puts it:

> Let us suppose that we are supplied with some unspecified means of solving number-theoretical [i.e., $\forall\exists$] problems; a kind of oracle as it were. ... With the help of the oracle we could form a new kind of machine

12 A relatively perspicuous representation in that form may be found in Davis et al. [6] p. 335.

> (call them *o*-machines), having as one of its fundamental processes that of solving a given number-theoretic problem.

He then showed that the problem of determining whether an *o*-machine terminates on any given input is an arithmetical problem not computable by any *o*-machine, and hence not solvable by the oracle itself. Turing did nothing further with the idea of *o*-machines, either in this paper or afterward. In 1944 Emil Post [20] took it as his basic notion for a theory of *degrees of unsolvability*, crediting Turing with the result that for any set of natural numbers there is another of higher degree of unsolvability. This transformed the notion of computability from an absolute notion into a relative one that would lead to entirely new developments and eventually to vastly generalized forms of recursion theory. Some of the basic ideas and results of the theory of effective reducibility of the membership problem for one set of numbers to another inaugurated by Turing and Post are explained in the article by Martin Davis in this issue of the *Notices*.

There are further interesting suggestions and asides in the thesis, such as consideration of possible constructive analogues of the Continuum Hypothesis. Finally (as also mentioned by Barry Cooper in his review article), it contained provocative speculations concerning intuition versus technical ingenuity in mathematical reasoning. The relevance, according to Turing is that:

> When we have an ordinal logic, we are in a position to prove number-theoretic theorems by the intuitive steps of recognizing [natural numbers as notations for ordinals]. . . . We want it to show quite clearly when a step makes use of intuition and when it is purely formal. . . . It must be beyond all reasonable doubt that the logic leads to correct results whenever the intuitive steps [i.e., recognition of ordinals] are correct.

This Turing had clearly accomplished with his formulation of the notion of ordinal logic and the construction of the particular S^* obtained by iterating consistency statements.

One reason that the reception of Turing's paper may have been so limited is that (no doubt at Church's behest) it was formulated in terms of the λ-calculus, which makes expressions for ordinals and formal systems very hard to understand. He could instead have followed Kleene, who wrote in his retrospective history [17]: "I myself, perhaps unduly influenced by rather chilly receptions from audiences around 1933–35 to disquisitions on λ-definability, chose, after general recursiveness had appeared, to put my work in that format. I cannot complain about my audiences after 1935."

ORDINAL LOGICS REDUX

The problems left open in Turing's thesis were attacked in my 1962 paper, "Transfinite recursive progressions of axiomatic theories" [7]. The title contains my rechristening of "ordinal logics" in order to give a more precise sense of the subject matter. In the interests of perspicuity and in order to explain what Turing had accomplished, I also recast all the notions in terms of general recursive functions and recursive notions for ordinals rather than the λ-calculus. Next I showed that Turing's progression based on iteration of consistency statements is not complete for true $\forall\exists$ statements, contrary to his hope. In fact, the same holds for the even stronger progression obtained by iterating adjunction to a system S of the *local reflection principle for S*. This is a scheme that formalizes, for each arithmetical sentence A, that if A is provable in S then A (is true). Then I showed that a progression $S^{(U)}$ based on the iteration of the *uniform reflection principle* is complete for all true arithmetical sentences. The latter principle is a scheme that formalizes, given S and a formula $A(x)$ that if $A(n)$ is provable in S for each n, then $\forall x A(x)$ (is true). One can also find a path P through O along which every true arithmetical sentence is provable in the progression $S^{(U)}$. On the other hand, invariance fails badly in the sense that there are paths P' through O for which there are true sentences in $\forall$-form not provable along that path, as shown in my paper with Spector [12]. The recent book *Inexhaustibility* [13] by Torkel Franzén contains an accessible introduction to [7], and his article [14] gives an interesting explanation (shorn of the offputting details) of what makes Turing's and my completeness results work.

The problem raised by Turing of recognizing which expressions (or numbers) are actually notations for ordinals is dealt with in part through the concept of *autonomous progressions of theories*, obtained by imposing a boot–strap procedure. That allows one to go to a system S_a *only* if one already has a proof in a previously accepted system S_b that $a \in O$ (or that a recursive ordering of order type corresponding to a is a well-ordering). Such progressions are not complete but have been used to propose characterizations of certain informal concepts of proof, such as that of finitist proof (Kreisel [18], [19]) and predicative proof (Feferman [8], [9]).

REFERENCES

[1] A. CHURCH, An unsolvable problem of elementary number theory, *Amer. J. of Math.* **58** (1936), 345–63. Reprinted in Davis [5].

[2] ——, A note on the Entscheidungsproblem, *J. Symbolic Logic* **1** (1936), 40–41; correction, ibid., 101–2. Reprinted in Davis [5].

[3] A. CHURCH and S. C. KLEENE, Formal definitions in the theory of ordinal numbers, *Fundamenta Mathematicae* **28** (1936), 11–21.

[4] A. M. COHEN and M. J. E. MAYHEW, On the difference $\pi(x) - \mathrm{li}(x)$, *Proc. London Math. Soc.* **18**(3) (1968), 691–713; reprinted in Turing [24].

[5] M. DAVIS, *The Undecidable. Basic Papers on Undecidable Propositions, Unsolvable Problems and Computable Functions*, Raven Press, Hewlett, NY (1965).

[6] M. DAVIS, YU. MATIJASEVIČ and J. ROBINSON, Hilbert's tenth problem. Diophantine equations: positive aspects of a negative solution, *Mathematical Developments Arising From Hilbert Problems*, (F. Browder, ed.) Amer. Math. Soc., Providence, RI (1976), 323–78.

[7] S. FEFERMAN, Transfinite recursive progressions of axiomatic theories, *J. Symbolic Logic* **27** (1962), 259–316.

[8] ——, Systems of predicative analysis, *J. Symbolic Logic* **29** (1964), 1–30.

[9] ——, Autonomous transfinite progressions and the extent of predicative mathematics, *Logic, Methodology and Philosophy of Science* III (B. van Rootselaar and J. F. Staal, eds.), North-Holland, Amsterdam (1968), 121–35.

[10] ——, Turing in the land of $O(z)$, in Herken [15], 113–47.

[11] ——, The impact of the incompleteness theorems on mathematics, *Notices Amer. Math. Soc.* **53** (April 2006), 434–39.

[12] S. FEFERMAN and C. SPECTOR, Incompleteness along paths in recursive progressions of theories, *J. Symbolic Logic* 27 (1962), 383–90.

[13] T. FRANZÉN, *Inexhaustibility. A non-exhaustive treatment*, Lecture Notes in Logic **28** (2004), Assoc. for Symbolic Logic, A. K. Peters, Ltd., Wellesley (distribs.).

[14] ——, Transfinite progressions: A second look at completeness, *Bull. Symbolic Logic* **10** (2004), 367–89.

[15] R. HERKEN (ed.), *The Universal Turing Machine. A Half-Century Survey*, Oxford University Press, Oxford (1988).

[16] A. HODGES, *Alan Turing: The Enigma*, Simon and Schuster, New York, 1983. New edition, Vintage, London, 1992.

[17] S. C. KLEENE, Origins of recursive function theory, *Ann. History of Computing* **3** (1981), 52–67.

[18] G. KREISEL, Ordinal logics and the characterization of informal concepts of proof, *Proc. International Congress of Mathematicians at Edinburgh* (1958), 289–99.

[19] ——, Principles of proof and ordinals implicit in given concepts, *Intuitionism and Proof Theory* (J. Myhill et al., eds.), North-Holland, Amsterdam (1970), 489–516.

[20] E. POST, Recursively enumerable sets and their decision problems, *Bull. Amer. Math. Soc.* **50** (1944), 284–316.

[21] A. M. TURING, On computable numbers, with an application to the Entscheidungsproblem, *Proc. London Math. Soc.* **42**(2) (1936–37), 230–65; correction, ibid. **43** (1937), 544–46. Reprinted in Davis [5] and Turing [25].

[22] ——, Computability and λ-definability, *J. Symbolic Logic* **2** (1937), 153–63. Reprinted in Davis [5] and Turing [25].

[23] ——, Systems of logic based on ordinals, *Proc. London Math. Soc.* (2) (1939), 161–228. Reprinted in Davis [5] and Turing [25].

[24] ——, *Pure Mathematics* (J. L. Britton, ed.), *Collected Works of A. M. Turing*, Elsevier Science Publishers, Amsterdam (1992).

[25] ——, *Mathematical Logic* (R. O. Gandy and C. E. M. Yates, eds.), *Collected Works of A. M. Turing*, Elsevier Science Publishers, Amsterdam (2001).

Notes on the Manuscript

> The thesis, which in October [1937] he had hoped to finish by Christmas, was delayed. "Church made a number of suggestions which resulted in the thesis being expanded to an appalling length." A clumsy typist himself, [Turing] engaged a professional, who in turn made a mess of it. It was eventually submitted on 17 May.
>
> —Andrew Hodges, *Alan Turing: the Enigma* (Princeton University Press, 2012)

In fact, the thesis is not a mess. Until the 1980s typists of mathematical texts routinely had to leave blank spaces where the mathematical symbols could be written in by hand. It's clearly Turing's handwriting (compare the capital A's and lowercase f's with his letter to Dean Eisenhart). There are only a few typos, where (e.g.) Turing crossed out "for" to write "of." The worst that could be said is that the typist did not always leave enough space for Turing to write in the formulas.

But pages 74–78 are typed by a less expert typist on a different typewriter: the type changes from elite to pica, and the typist does not strike all the letters with equal force. Did Turing type pages 74–78 himself?

In May 1938, Turing submitted "Systems of Logic Based on Ordinals" for publication in the *Proceedings of the London Mathematical Society.* It was refereed in June 1938 and appeared in 1939. The printed version adheres very closely to the manuscript reproduced here. Where there are differences in the refereed article that change the import of the mathematics, they are noted with an arrowhead (➤) in the margin of the manuscript, and the corresponding text from the *Proc. LMS* appears here.

11. (b) $\varphi(x_1, \ldots, x_n) = f(x_2, \ldots, x_n)$;

12. $\varphi(x, y) = \theta(x) + \alpha(x, y)$,

12. $\theta_1(0) = 3$,

$\theta_1(x+1) = 2^{(1+\bar{\omega}_2(\theta_1(x)))\sigma(\varphi(\bar{\omega}_3(\theta_1(x))-1,\, \bar{\omega}_2(\theta_1(x))))}\, 3^{\bar{\omega}_3(\theta_1(x))+1-\sigma(\varphi(\bar{\omega}_3(\theta_1(x))-1,\, \bar{\omega}_2(\theta_1(x))))}$,

13. $\varphi(x, y)$ are primitive recursive functions. Without loss of generality, we may suppose that the functions φ, ψ take only the values 0, 1. Then, if we define $\rho(x)$ by the equation (3.1) and

$\rho(0) = \psi(0)(1 - \theta(0))$,

$\rho(x+1) = 1 - (1 - \rho(x))\sigma[1 + \theta(x) - \psi\{\bar{\omega}_2(\theta_1(x))\}]$

15. If x is the G.R. of 2 (*i.e.* if x is $2^3 . 3^{10} . 5 . 7^3 . 11^{28} . 13 . 17 . 19^{10} . 23^2 . 29 . 31 . 37^{10} . 41^2 . 43 . 47^{28} . 53^2 . 59^2 . 61^2 . 67^2$) and let $c(x)$ be 1 otherwise.

19. o-machine whose description number is $r(n)$. This o-machine is circle free

27. Let Nm be a W.F.F. which enumerates all formulae with normal forms and no free variables.

28. $(D')\{(\exists x)(D'(x))\ \&\ (x)(D'(x) \supset D(x)) \supset (\exists z)(y)[D'(z)\ \&\ (D'(y) \supset G(z, y) \vee z = y)]\}$. (7.2)

34. (iv) If **A**, **B**, **C** are C-K ordinal formulae and **B**<**A**, **C**<**A**, then either **B**<**C**, **C**<**B**, or **B** conv **C**.

34. for which $\mathbf{B}_r < \mathbf{B}_{r-1} < \mathbf{A}$ for each r.

35. Suc (λufx . **B**) conv Suc (λufx . **B**$'$) and λufx . u(**R**) conv λufx . u(**R**$'$),

36. are convertible to the forms λufx . **B**, λufx . **B**$'$; but

38. (for some n')

39. the conditions $\lambda ufx . \mathbf{R}(\mathbf{n}) < \lambda ufx . \mathbf{R}(S(\mathbf{n}))$ in (B).

42. Sum $\rightarrow \lambda ww'\, pq$. Bd(w, w', Hf(p), Hf(q), Al(p, Al(q,w'(Hf(p), Hf(q)), 1), Al(S(q), w(Hf(p), Hf(q)), 2))),

43. sequence of ordinal formulae representing all the ordinals less than α without repetitions other than repetitions of the ordinal 0.

46. To prove this we shall show that to each C-K ordinal formula **A** there corresponds a unique system C[**A**] such that:

(i) $\mathbf{A}(\Theta, \mathbf{K}, \mathbf{m}_{C_0})$ conv $\mathbf{m}_{\mathrm{C[A]}}$,

52. $\mathfrak{C}_r[y_0] \supset (\exists x_0)((\mathrm{Db}[x_0, y_0] . \mathfrak{C}_r[y_0]) \vee (\mathrm{Db}[fx_0, fy_0] . \mathfrak{U}_{r+1}[y_0]))$

and

$\mathfrak{U}_{r+1}[y_0] \supset (\exists x_0)((\mathrm{Db}[x_0, y_0] . \mathfrak{C}_r[y_0]) \vee (\mathrm{Db}[fx_0, fy_0] . \mathfrak{U}_{r+1}[y_0]))$.

56. $(c)\ m = 2p - 1, n = 2q - 1$, and $\Omega(\mathbf{p}, \mathbf{q})$ conv 2.
58. $\mathrm{Ai} \rightarrow \lambda kw \,.\, \Gamma(\lambda ra \,.\, \delta(4, \delta(2, k(w, V(\mathrm{Nm}(r)))) + \delta(2, \mathrm{Nm}(r, a)))),$
59. $\mathrm{Ai}(\Lambda, \Omega_{V(\mathrm{L})}, \mathbf{B})$ is convertible to
$\Gamma(\lambda ra \,.\, \delta(4, \delta(2, \Lambda(\Omega_{V(\mathrm{L})}, \mathrm{V}(\mathrm{Nm}(r)))) + \delta(2, \mathrm{Nm}(r, a))), \mathbf{B}).$
64. Now let us turn to Λ_H.
64. if G is provable in P_Ω it is provable in $P_{\Omega'}$. Λ_H is invariant.
65. but, if $\mathbf{A}(\mathbf{c})$ is not convertible to 2, then
68. This means that $\mathbf{M}(\mathbf{n})$ is convertible to 2.
70. type 3 being the highest necessary.
71. the ordinals less than ω^2 take the place of
72. belong to the extent of $\mathbf{\Lambda}(\mathbf{G}(\mathbf{\Phi}(\lambda r \,.\, \mathrm{Hg}(\mathbf{A}, r, \mathbf{E})))),$
78. Then we have as an axiom in P . . .
and we can prove in P^{A}
96. $g(r_1, r_2, \ldots, r_p)$ has the value p.

SYSTEMS OF LOGIC BASED ON ORDINALS

A. M. Turing

(A dissertation presented to the faculty of Princeton University in candidacy for the degree of Doctor of Philosophy)

Recommended by the
Department of Mathematics
for acceptance.

May 1938.

CONTENTS

The well known theorem of Gödel shows that every system of logic is in a certain sense incomplete, but at the same time it indicates means whereby from a system L of logic a more complete system L' may be obtained. By repeating the process we get a sequence $L, L_1 = L', L_2 = L_1', L_3 = L_2'$,... of logics each more complete than the preceding. A logic L_ω may then be constructed in which the provable theorems are the totality of theorems provable with the help of the logics L, L_1, L_2 ,... We may then form $L_{2\omega}$ related to L_ω in the same way as L_ω was related to L. Proceeding in this way we can associate a system of logic with any given constructive ordinal.[1] It may be asked whether a sequence of logics of this kind is complete in the sense that to any problem A there corresponds an ordinal α such that A is solvable by means of the logic L_α. I propose to investigate this problem in a rather more general case, and to give some other examples of ways in which systems of logic may be associated with constructive ordinals.

[1] The situation is not quite so simple as is suggested by this crude argument. See pages 44-48.

1. <u>The calculus of conversion</u>. <u>Gödel representations</u>.

It will be convenient to be able to use the 'conversion calculus' of Church for the description of functions and some other purposes. This will make greater clarity and simplicity of expression possible. I shall give a short account of this calculus. For more detailed descriptions see Church [3], [2], Kleene [1],

Church and Rosser [1].

The formulae of the calculus are formed from the symbols $\{$, $\}$ $($, $)$, $[$, $]$, λ, δ, and an infinite list of others called variables; we shall take for our infinite list $a, b, \ldots, z, x', x'', \ldots$ Certain finite sequences of such symbols are called well-formed formulae (abbreviated to W.F.F.); we shall define this class inductively, and simultaneously define the free and the bound variables of a W.F.F. Any variable is a W.F.F.; it is its only free variable, and it has no bound variables. δ is a W.F.F. and has no free or bound variables. If $\underline{M}$ and $\underline{N}$ are W.F.F. then $\{\underline{M}\}(\underline{N})$ is a W.F.F. whose free variables are the free variables of $\underline{M}$ together with the free variables of $\underline{N}$, and whose bound variables are the bound variables of $\underline{M}$ together with those of $\underline{N}$. If $\underline{M}$ is a W.F.F. and $\underline{V}$ one of its free variables, then $\lambda\underline{V}[\underline{M}]$ is a W.F.F. whose free variables are those of $\underline{M}$ with the exception of $\underline{V}$, and whose bound variables are those of $\underline{M}$ together with $\underline{V}$. No sequence of symbols is a W.F.F. except in consequence of these three statements.

In metamathematical statements we shall use underlined letters to stand for variable or undetermined formulae, as was done in the last paragraph, and in future such letters will stand for well-formed formulae unless otherwise stated. Small letters underlined will stand for formulae representing undetermined positive integers (see below).

A W.F.F. is said to be in normal form if it has no parts of the form $\{\lambda\underline{V}[\underline{M}]\}(\underline{N})$ and none of the form $\{\{\delta\}(\underline{M})\}(\underline{N})$ where $\underline{M}$ and $\underline{N}$ have no free variables.

We say that one W.F.F. is <u>immediately convertible</u> into another if it is obtained from it either by

(i) Replacing one occurrence of a well-formed part $\lambda \underline{V}[\underline{M}]$ by $\lambda \underline{U}[\underline{N}]$, where the variable $\underline{U}$ does not occur in $\underline{M}$, and $\underline{N}$ is obtained from $\underline{M}$ by replacing the variable $\underline{V}$ by $\underline{U}$ throughout.

(ii) Replacing a well-formed part $\{\lambda \underline{V}[\underline{M}]\}(\underline{N})$ by the formula which is obtained from $\underline{M}$ by replacing $\underline{V}$ by $\underline{N}$ throughout, provided that the bound variables of $\underline{M}$ are distinct both from $\underline{V}$ and from the free variables of $\underline{N}$.

(iii) The converse process of ii.

(iv) Replacing a well-formed part $\{\{S\}(\underline{M})\}(\underline{M})$ by $\lambda f[\lambda x[\{f\}(\{f\}(x))]]$ if $\underline{M}$ is in normal form and has no free variables.

(v) Replacing a well-formed part $\{\{S\}(\underline{M})\}(\underline{N})$ by $\lambda f[\lambda x[\{f\}(x)]]$ if $\underline{M}$ and $\underline{N}$ are in normal form and not transformable into one another by repeated application of i, and have no free variables.

(vi) The converse process of iv.

(vii) The converse process of v.

These rules could have been expressed in such a way that in no case could there be any doubt as to the admissibility or the result of the transformation (in particular this can be done in the case of process v.).

A formula $\underline{A}$ is said to be <u>convertible</u> into another $\underline{B}$ (abbreviated to '$\underline{A}$ conv $\underline{B}$') if there is an finite chain of immediate conversions leading from one formula to the other. It is easily seen that the relation of convertibility is an equivalence relation, i.e. it is symmetric, transitive and reflexive.

Since the formulae are liable to be very lengthy we need means for abbreviating them. If we wish to introduce a particular letter as an abbreviation for a particular lengthy formula we shall write the letter followed by '$\rightarrow$' and then by the formula, thus

$$\underline{I} \rightarrow \lambda x[x]$$

indicates that $\underline{I}$ is an abbreviation for $\lambda x[x]$. We shall also use the arrow in less sharply defined senses, but never so as to cause any real confusion. In these cases the meaning of the arrow may be rendered by the words 'stands for'.

If a formula $\underline{F}$ is, or is represented by, a single symbol we abbreviate $\{\underline{F}\}(\underline{X})$ to $\underline{F}(\underline{X})$. A formula $\{\{\underline{F}\}(\underline{X})\}(\underline{Y})$ may be abbreviated to $\{\underline{F}\}(\underline{X}, \underline{Y})$, or to $\underline{F}(\underline{X}, \underline{Y})$ if $\underline{F}$ is, or is represented by, a single symbol. Similarly for $\{\{\{\underline{F}\}(\underline{X})\}(\underline{Y})\}(\underline{Z})$ etc. A formula $\lambda \underline{V}_1[\lambda \underline{V}_2 \ldots [\lambda \underline{V}_r[\underline{M}]] \ldots]$ may be abbreviated to $\lambda \underline{V}_1 \underline{V}_2 \ldots \underline{V}_r . \underline{M}$.

We have not as yet assigned any meanings to our formulae, and we do not intend to do so in general. An exception may be made for the case of the positive integers which are very conveniently represented by the formulae $\lambda fx . f(x)$, $\lambda fx . f(f(x))$, ... In fact we introduce the abbreviations

$$1 \rightarrow \lambda fx . f(x)$$
$$2 \rightarrow \lambda fx . f(f(x))$$
$$3 \rightarrow \lambda fx . f(f(f(x)))$$

etc.

and also say for example that $\lambda fx \,.\, f(f(x))$ (in full $\lambda f[\lambda x[\{f\}(\{f\}(x))]]$ represents the positive integer 2. Later we shall allow certain formulae to represent ordinals, but otherwise we leave them without explicit meaning; an implicit meaning may be suggested by the abbreviations used. In any case where any meaning is assigned to formulae it is desirable that the meaning be invariant under conversion. Our definitions of the positive integers do not violate this requirement, as it may be proved that no two formulae representing different positive integers are convertible into one another.

In connection with the positive integers we introduce the abbreviation

$$S \rightarrow \lambda ufx \,.\, f(u(f,x))$$

This formula has the property that if $\underline{n}$ represents a positive integer $S(\underline{n})$ is convertible to a formula representing its successor.[2]

---

[2] This follows from (A) below.

Formulae representing undetermined positive integers will be represented by small letters underlined, and we shall adopt once for all the convention that if a letter, n say, stands for a positive integer, then the same letter underlined, $\underline{n}$, stands for the formula representing the positive integer. When no confusion arises from doing so we shall omit to distinguish between an integer and the formula which represents it.

Suppose $f(n)$ is a function of positive integers taking positive integers as values, and that there is a W.F.F. $\underline{F}$ not containing δ such that for each positive integer n, $\underline{F}(\underline{n})$ is convertible to the formula representing $f(n)$. We shall then say that $f(n)$ is λ -definable or formally definable, and that $\underline{F}$ formally defines $f(n)$. Similar conventions are used for functions of more than one variable. The sum function is for instance formally defined by $\lambda abfx.a(f,b(f,x))$; in fact for any positive integers m, n, p for which $m+n=p$ we have

$\{\lambda abfx.a(f,b(f,x))\}(\underline{m},\underline{n})$ conv $\underline{p}$

In order to emphasize this relation we introduce the abbreviation

$$\underline{X}+\underline{Y} \rightarrow \{\lambda abfx.a(f,b(f,x))\}(\underline{X},\underline{Y})$$

and will use similar notations for sums of three or more terms, products etc.

For any W.F.F. $\underline{G}$ we shall say that $\underline{G}$ enumerates the sequence $\underline{G}(1)$, $\underline{G}(2)$, . . . , and any other sequence whose terms are convertible to those of this sequence.

When a formula is convertible to another which is in normal form the second is described as a normal form of the first, which is then said to have a normal form. I quote here some of the more important theorems concerning normal forms.

(A) If a formula has two normal forms they are convertible into one another by the use of (i) alone. (Church and Rosser [1], 479, 481).

(B) If a formula has a normal form then every well-formed part of it has a normal form. (Church and Rosser [1], 480-481).

(C) There is (demonstrably) no process whereby one can

tell of a formula whether it has a normal form. (Church [3], 360, Theorem XVIII.)

We often need to be able to describe formulae by means of positive integers. The method used here is due to Gödel (Gödel [1]). To each single symbol S of the calculus we assign an integer $r[S]$ as in the table below.

S	{, (or [	},) or]	λ	δ	a	...	z	x'	x''	x'''	...
$r[S]$	1	2	3	4	5	...	30	31	32	33	...

If $S_1 S_2 \ldots S_k$ is a sequence of symbols then $2^{r[S_1]} 3^{r[S_2]} \ldots p_k^{r[S_k]}$ (where p_k is the k th prime number) is called the <u>Gödel representation</u> (G.R.) of that sequence of symbols. No two W.F.F. have the same G.R.

Two theorems on G.R. of W.F.F. are quoted here.

(D) There is a W.F.F. $form$ such that if a is the G.R. of a W.F.F. $\underline{A}$ without free variables then $form(\underline{a})$ conv $\underline{A}$. (This follows from a similar theorem to be found in Church [3], 53-66. Metads are used there in place of G.R.)

(E) There is a W.F.F. Gr such that if $\underline{A}$ is a W.F.F. with a normal form without free variables, then $Gr(\underline{A})$ conv $\underline{a}$, where a is the G.R. of a normal form of $\underline{A}$. (Church [3], 53-66, as (D)).

2. Effective calculability. Abbreviation of treatment.

A function is said to be 'effectively calculable' if its values can be found by some purely mechanical process. Although it is fairly easy to get an intuitive grasp of this idea it is nevertheless desirable to have some more definite, mathematically expressible definition. Such a definition was first given by Gödel at Princeton in 1934 (Gödel [2], 26) following in part an unpublished suggestion of Herbrand, and has since been developed by Kleene (Kleene [2]). We shall not be concerned much here with this particular definition. Another definition of effective calculability has been given by Church (Church [3], 356-358) who identifies it with λ-definability. The author has recently suggested a definition corresponding more closely to the intuitive idea (Turing [1], see also Post [1]). It was said above "a function is effectively calculable if its values can be found by some purely mechanical process." We may take this statement literally, understanding by a purely mechanical process one which could be carried out by a machine. It is possible to give a mathematical description, in a certain normal form, of the structures of these machines. The development of these ideas leads to the author's definition of a computable function, and an identification of computability[3] with effective calculability.

---

[3] We shall use the expression 'computable function' to mean a function calculable by a machine, and let 'effectively calculable' refer to the intuitive idea without particular identification with any one of these definitions. We do not restrict the values taken by a computable function to be natural numbers; we may for instance have computable propositional functions.

It is not difficult through somewhat laborious, to prove these three definitions equivalent (Kleene [3], Turing [2]).

In the present paper we shall make considerable use of Church's identification of effective calculability with λ-definability, or, what comes to the same, of the identification with computability and one of the equivalence theorems. In most cases where we have to deal with an effectively calculable function we shall introduce the corresponding W.F.F. with some such phrase as "the function f is effectively calculable, let F be a formula λ-defining it" or "let F be a formula such that $F(\underline{n})$ is convertible to . . . whenever $\underline{n}$ represents a positive integer". In such cases there is no difficulty in seeing how a machine could in principle be designed to calculate the values of the function concerned, and assuming this done the equivalence theorem can be applied. A statement as to what the formula F actually is may be omitted. We may introduce immediately on this basis a W.F.F. $\mathfrak{D}$ with the property that $\mathfrak{D}(\underline{m}, \underline{n})$ conv $\underline{r}$ if r is the greatest positive integer for which m^r divides n, if any, and is 1 if there is none. We also introduce $\mathfrak{Dt}$ with the properties

$$\mathfrak{Dt}(\underline{n}, \underline{n}) \text{ conv } 3$$

$$\mathfrak{Dt}(\underline{n}+\underline{m}, \underline{n}) \text{ conv } 2$$

$$\mathfrak{Dt}(\underline{n}, \underline{n}+\underline{m}) \text{ conv } 1$$

There is another point to be made clear in connection with the point of view we are adopting. It is intended that all proofs that are given should be regarded no more critically than proofs in classical analysis. The subject matter, roughly speaking, is constructive systems of logic, but as the purpose is directed towards choosing a particular constructive system of logic for practical use; an attempt at this stage to put our theorems into constructive form would be putting the cart before the horse.

Those computable functions which take only the values 0 and 1 are of particular importance since they determine and are determined by computable properties, as may be seen by replacing '0' and '1' by 'true' and 'false'. But besides this type of property we may have to consider a different type, which is, roughly speaking, less constructive than the computable properties, but more so than the general predicates of classical mathematics. Suppose we have a computable function of the natural numbers taking natural numbers as values, then corresponding to this function there is the property of being a value of the function. Such a property we shall describe as 'axiomatic'; the reason for using this term is that it is possible to define such a property by giving a set of axioms, the property to hold for a given argument if and only if it is possible to deduce that it holds from the axioms.

Axiomatic properties may also be characterized in this way. A property ψ of positive integers is axiomatic if and only if there is a computable property φ of two positive integers, such that $\psi(x)$ is true if and only if there is a positive integer y such that $\varphi(x, y)$ is true. Or again ψ is axiomatic if and only if there is a W.F.F. $\underline{F}$ such that $\psi(n)$ is true if and only if $\underline{F}(\underline{n})$ conv 2.

3. Number theoretic theorems

By a <u>number theoretic theorem</u>[4] we shall mean a theorem of the

[4] I believe there is no generally accepted meaning for this term, but it should be noticed that we are using it in a rather restricted sense. The most generally accepted meaning is probably this: suppose we take an arbitrary formula of the function calculus of first order and replace the function variables by primitive recursive relations. The resulting formula represents a typical number theoretic theorem in this (more general) sense.

form '$\theta(x)$ vanishes for infinitely many natural numbers x', where $\theta(x)$ is a primitive recursive[5] function.

[5] Primitive recursive functions of natural numbers are defined inductively as follows: Suppose $f(x_1, \ldots, x_{n-1})$, $g(x_1, \ldots, x_n)$, $h(x_1, \ldots, x_{n+1})$ are primitive recursive then $\varphi(x_1, \ldots, x_n)$ is primitive recursive if it is defined by one of the sets of equations (a) - (e).

(a) $\varphi(x_1, \ldots, x_n) = h(x_1, \ldots, x_{m-1}, g(x_1, \ldots, x_n), x_{m+1}, \ldots, x_{n-1}, x_n)$, $(1 \leq m \leq n)$

(b) $\varphi(x_1, \ldots, x_n) = f(x_1, \ldots, x_{n-1})$

(c) $\varphi(x_1) = a$, where $n = 1$ and a is some particular natural number.

(d) $\varphi(x_1) = x_1 + 1$ $(n = 1)$

(e) $\varphi(x_1, \ldots, x_{n-1}, 0) = f(x_1, \ldots, x_{n-1})$
$\varphi(x_1, \ldots, x_{n-1}, x_n + 1) = h(x_1, \ldots, x_n, \varphi(x_1, \ldots, x_n))$

The class of primitive recursive function is more restricted than the computable functions, but has the advantage that there is a process whereby one can tell of a set of equations whether if defines a primitive recursive function in the manner described above.

If $\varphi(x_1, \ldots, x_n)$ is primitive recursive then $\varphi(x_1, \ldots, x_n) = 0$ is described as a primitive recursive relation between $x_1, \ldots, x_n$.

We shall say that a problem is number theoretic if it has been shown that any solution of the problem may be put in the form of a proof of one or more number theoretic theorems. More accurately we may say that a class of problems is number theoretic if the solution of any one of them can be transformed (by a uniform process) into the form of proofs of number theoretic theorems.

I shall now draw a few consequences from the definition of 'number theoretic theorems', and in section 5 will try to justify confining our considerations to this type of problem.

An alternative form for 'number theoretic theorems' is 'for each natural number x there exists a natural number y such that $\varphi(x, y)$ vanishes', where $\varphi(x, y)$ is primitive recursive and conversely. In other words, there is a rule whereby given the function $\theta(x)$ we can find a function $\varphi(x, y)$, or given $\varphi(x, y)$ we can find a function $\theta(x)$, so that '$\theta(x)$ vanishes infinitely often' is a necessary and sufficient condition for 'for each x there is y so that $\varphi(x, y) = 0$'. In fact given $\theta(x)$ we define

$$\varphi(x, y) = \theta(y) + \alpha(x, y)$$

where $\alpha(x, y)$ is the (primitive recursive) function with the properties

$$\begin{aligned} \alpha(x, y) &= 1 \quad (y \leq x) \\ &= 0 \quad (y > x) \end{aligned}$$

If on the other hand we are given $\varphi(x, y)$ we define $\theta(x)$ by the equations

$$\begin{aligned} \theta_1(0) &= 3 \\ \theta_1(x+1) &= 3 \cdot \tfrac{2}{3}(\theta_1(x))^{\sigma(\varphi(\vartheta_3(\theta_1(x)) - 1,\, \vartheta_3(\theta_1(x))))} \\ \theta(x) &= \varphi(\vartheta_3(\theta_1(x)) - 1,\, \vartheta_2(\theta_1(x))) \end{aligned}$$

where $\vartheta_r(x)$ is to be defined so as to mean 'the largest s for which r^s divides x ' and $\frac{2}{3}x$ to be defined primitive recursively so as to have its usual meaning if x is a multiple of 3. The function $\sigma(x)$ is to be defined by the equations $\sigma(0) = 0$, $\sigma(x+1) = 1$. It is easily verified that the functions so defined have the desired properties.

We shall now show that questions as to the truth of statements of form 'does $f(x)$ vanish identically', where $f(x)$ is a computable function, can be reduced to questions as to the truth of number theoretic theorems. It is understood that in each case the rule for the calculation of $f(x)$ is given and that one is satisfied that this rule is valid, i.e. that the machine which should calculate $f(x)$ is circle free (Turing [1], 233). The function $f(x)$ being computable is general recursive in the Herbrand-Gödel sense, and therefore by a general theorem due to Kleene[6] is expressible in the form

[6] Kleene [3], 727. This result is really superfluous for our purpose, as the proof that every computable function is general recursive proceeds by showing that these functions are of form (3.2). (Turing [2], 161).

$$\psi(\epsilon y[\varphi(x,y) = 0]) \tag{3.2}$$

where $\epsilon y[\mathfrak{A}(y)]$ means 'the least y for which $\mathfrak{A}(y)$ is true' and $\psi(y)$ and $\varphi(x,y)$ are primitive recursive functions. Then if we define $\rho(x)$ by the equations (3.1) and

$$\rho(x) = \varphi(\vartheta_3(\theta_1(x)) - 1,\ \vartheta_2(\theta_1(x))) + \psi(\vartheta_2(\theta_1(x)))$$

it will be seen that $f(x)$ vanishes identically if and only if $\rho(x)$ vanishes for infinitely many values of x.

The converse of this result is not quite true. We cannot say that the question as to the truth of any number theoretic theorem is reducible to a question as to whether a corresponding computable function vanishes identically; we should have rather to say that it is reducible to the problem as to whether a certain machine is circle free and calculates an identically vanishing function. But more is true: every number theoretic theorem is equivalent to the statement that a corresponding machine is circle free. The behavior of the machine may be described roughly as follows: the machine is one for the calculation of the primitive recursive function $\theta(x)$ of the number theoretic problem, except that the results of the calculation are first arranged in a form in which the figures 0 and 1 do not occur, and the machine is then modified so that whenever it has been found that the function vanishes for some value of the argument, then 0 is printed. The machine is circle free if and only if an infinity of these figures are printed, i.e. if and only if $\theta(x)$ vanishes for infinitely many values of the argument. That, on the other hand, questions as to circle freedom may be reduced to questions of the truth of number theoretic theorems follows from the fact that $\theta(x)$ is primitive recursive when it is defined to have the value 0 if a certain machine $\mathcal{M}$ prints 0 or 1 in its $(x+1)$ th complete configuration, and to have the value 1 otherwise.

The conversion calculus provides another normal form for the number theoretic theorems, and the one we shall find the most convenient to use. Every number theoretic theorem is equivalent to a statement of the form '$\underline{A}(\underline{n})$ is convertible to 2 for every W.F.F. $\underline{n}$ representing a positive integer', $\underline{A}$ being a W.F.F. determined by the theorem; the property of $\underline{A}$ here asserted will be described briefly as '$\underline{A}$ is dual'. Conversely such statements are reducible to number theoretic theorems. The first half of this assertion follows from our results for computable functions, or directly in this way. Since $\theta(x-1)+2$ is primitive recursive it is formally definable, by means of a formula $\underline{G}$ let us say. Now there is (Kleene [1], 232) a W.F.F. $\mathfrak{P}$ with the property that if $\underline{T}(\underline{r})$ is convertible to a formula representing a positive integer for each positive integer r, then $\mathfrak{P}(\underline{T},\underline{n})$ is convertible to $\underline{s}$ where s is the nth positive integer t (if there is one) for which $\underline{T}(\underline{t})$ conv 2; if $\underline{T}(\underline{t})$ conv 2 for less than n values of t then $\mathfrak{P}(\underline{T},\underline{n})$ has no normal form. The formula $\underline{G}(\mathfrak{P}(\underline{G},\underline{n}))$ will therefore be convertible to 2 if and only if $\theta(x)$ vanishes for at least n values of x, and will be convertible to 2 for every positive integer x if and only if $\theta(x)$ vanishes infinitely often. To prove the second half of the assertion we take Gödel representations for the formulae of the conversion calculus. Let $c(x)$ be 0 if x is the G. R. of 2 (i.e. if x is $2^3.3^{10}.5.7^3.11^{28}.13.17.19^{10}.23^2.29.31.37^{28}.41^2.43.47^{28}.53^2.59^2.61^2.67^2$) and otherwise be 1. Take an enumeration of the G. R. of the formulae into which $\underline{A}(\underline{m})$ is convertible: let $a(m,n)$ be the nth number in the enumeration. We can arrange the enumeration so that $a(m,n)$ is primitive recursive. Now the statement that $\underline{A}(\underline{m})$

is convertible to 2 for every positive integer m is equivalent to the statement that for each positive integer m there is a positive integer n such that $c(a(m,n)) = 0$, and this is number theoretic.

It is easy to show that a number of unsolved problems such as the problem as to the truth of Fermat's last theorem are number theoretic. There are, however, also problems of analysis which are number theoretic. The Riemann hypothesis gives us an example of this. We denote by $\zeta(s)$ the function defined for $\mathcal{R}s = \sigma > 1$ by the series $\sum_{n=1}^{\infty} \frac{1}{n^s}$ and over the rest of the complex plane with the exception of the point $s = 1$ by analytic continuation. The Riemann hypothesis asserts that this function does not vanish in the domain $\sigma > \frac{1}{2}$. It is easily shown that this is equivalent to saying that it does not vanish for $2 > \sigma > \frac{1}{2}$, $\mathcal{R}s = t > 2$ i.e. that it does not vanish inside any rectangle $2 > \sigma > \frac{1}{2} + \frac{1}{T}$, $T > t > 2$ where T is an integer greater than 2. Now the function satisfies the inequalities

$$\left.\begin{aligned} &\left| \zeta(s) - \sum_{1}^{n} n^{-s} - \frac{N^{1-s}}{s-1} \right| < 2t(N-2)^{-\frac{1}{2}} \\ &\left| \zeta(s) - \zeta(s') \right| < |s-s'| . 60t \end{aligned}\right\} \quad \begin{aligned} &2 < \sigma < \tfrac{1}{2}, \ t \geqslant 2 \\ &2 < \sigma' < \tfrac{1}{2}, \ t' \geqslant 2 \end{aligned}$$

and we can define a primitive recursive function $\xi(l, l', m, m', N, M)$ such that

$$\left| \xi(l, l', m, m', N, M) - M \left| \sum_{1}^{N} n^{-s} + \frac{N^{1-s}}{s-1} \right| \right| < 2 \quad \left(s = \frac{l}{l'} + i\frac{m}{m'}\right)$$

and therefore if we put

$$\xi(l, M, m, M, M^2+2, M) = X(l, m, M)$$

we shall have

$$\left|\zeta\left(\frac{l+\vartheta}{M}+i\,\frac{m+\vartheta'}{M}\right)\right| \geqslant \frac{X(l,m,M)-122T}{M}$$

$$\frac{1}{2}+\frac{1}{T}\leq\frac{l-1}{M}<\frac{l+1}{M}<2-\frac{1}{M},\ 2<\frac{m-1}{M}<\frac{m+1}{M}<T,\ -1<\vartheta<1,\ -1<\vartheta'<$$

if we define $B(M,T)$ to be the smallest value of $X(l,m,M)$ for which $\frac{1}{2}+\frac{1}{T}+\frac{1}{M}\leq\frac{l}{M}<2-\frac{1}{M},\ 2+\frac{1}{M}<\frac{m}{M}<T-\frac{1}{M}$, then the Riemann hypothesis is true if for each T there is M satisfying $B(M,T)>122T$. If on the other hand there is T such that for all M, $B(M,T)\leq 122T$, the Riemann hypothesis is false; for let l_M, m_M be such that $X(l_M, m_M, M)\leq 122T$ then $\left|\zeta\left(\frac{l_M+im_M}{M}\right)\right|\leq\frac{244T}{M}$

Now if a is a condensation point of the sequence $\frac{l_M+im_M}{M}$ then since $\zeta(s)$ is continuous except at $s=1$ we must have $\zeta(a)=0$ implying the falsity of the Riemann hypothesis. Thus we have reduced the problem to the question as to whether for each T there is M for which $B(M,T)>122T$. $B(M,T)$

is primitive recursive, and the problem is therefore number theoretic.

4. A type of problem which is not number theoretic.[7]

[7] Compare Rosser [1].

Let us suppose that we are supplied with some unspecified means of solving number theoretic problems; a kind of oracle as it were. We will not go any further into the nature of this oracle than to say that it cannot be a machine. With the help of the oracle we could form a new kind of machine (call them o-machines), having as one of its fundamental processes that of solving a given number theoretic problem. More definitely these machines are to behave in this way. The moves of the machine are determined as usual by a table except in the case of moves from a certain internal configuration $\mathfrak{o}$. If the machine is in the internal configuration $\mathfrak{o}$ and if the sequence of symbols marked with $\mathfrak{l}$ is then the well formed[8] formula $\underline{A}$, then the machine goes into the internal

[8] Without real loss of generality we may suppose that $\underline{A}$ is always well formed.

configuration $\mathfrak{p}$ or $\mathfrak{t}$ according as it is or is not true that $\underline{A}$ is dual. The decision as to which is the case is referred to the oracle.

These machines may be described by tables of the same kind as used for the description of a-machines, there being no entries, however, for the internal configuration $\mathfrak{o}$. We obtain description numbers from these tables in the same way as before. If we make the convention that in assigning numbers to internal configurations $\mathfrak{o}$, $\mathfrak{p}$, $\mathfrak{t}$ are always to be q_2, q_3, q_4, then the description numbers determine the behaviour of the machines uniquely.

Given any one of these machines we may ask ourselves the question whether or not it prints an infinity of figures 0 or 1; I assert that this class of problems is not number theoretic. In view of the definition of 'number theoretic problem' this means to say that it is not possible to construct an o-machine which when supplied[9] with the description of any other o-machine will determine whether that machine is o-circle free. The argument may be taken over directly from Turing [1], p. 8. We say that a number is o-satisfactory if it is the description number of an o-circle free machine. Then if there is an o-machine which will determine of any integer whether it is o-satisfactory then there is also an o-machine to calculate the values of the function $1 - \varphi_n(n)$. Let $r(n)$ be the nth o-satisfactory number and let $\varphi_n(m)$ be the mth figure printed by the o-machine whose description number is n. This o-machine is circle free and there is therefore an o-satisfactory number K such that $\varphi_K(n) = 1 - \varphi_n(n)$ all n. Puting $n = K$ yields a contradiction. This completes the proof that problems of circle freedom of o-machines are not number theoretic.

Propositions of the form that an o-machine is o-circle free can always be put in the form of propositions obtained from formulae of the functional calculus of first order by replacing <u>some</u> of the functional variables by primitive recursive relations. Compare footnote[6].

---

[9] Compare Turing [1], § 6,7.

5. Syntactical theorems as number theoretic theorems.

I shall mention a property of number theoretic theorems which suggests that there is reason for regarding them as of particular importance.

Suppose that we have some axiomatic system of a purely formal nature. We do not interest ourselves at all in interpretations for the formulae of this system. They are to be regarded as of interest for themselves. An example of what is in mind is afforded by the conversion calculus (§1). Every sequence of symbols '$\underline{A}$ conv $\underline{B}$' where $\underline{A}$ and $\underline{B}$ are well formed formulae, is a formula of the axiomatic system and is provable if the W.F.F. $\underline{A}$ is convertible to $\underline{B}$. The rules of conversion give us the rules of procedure in this axiomatic system.

Now consider a new rule of procedure which is reputed to yield only formulae provable in the original sense. We may ask ourselves whether such a rule is valid. The statement that such a rule is valid would be number theoretic. To prove this let us take Gödel representations for the formulae, and an enumeration of the provable formulae; let $\varphi(r)$ be the G. R. of the r th formula in the enumeration. We may suppose $\varphi(r)$ is primitive recursive if we do not mind repetitions in the enumeration. Let $\psi(r)$ be the G. R. of the r th formula obtained by the new rule, then the statement that this new rule is valid is equivalent to the assertion of

$$(r)(\exists s)[\psi(r) = \varphi(s)]$$

(the domain of individuals being the natural numbers). It has been shown in § 3 that such statements are number theoretic.

It might plausibly be argued that all theorems of mathematics which have any significance when taken alone, are in effect syntactical theorems of this kind, stating the validity of certain 'derived rules' of procedure. Without going so far as this I should say that theorems of this kind have an importance which makes it worth while to give them special consideration.

6. Logic formulae

We shall call a formula $\underline{L}$ a logic formula (or, if it is clear that we are speaking of a W.F.F., simply a logic) if it has the property that if $\underline{A}$ is a formula such that $\underline{L}(\underline{A})$ conv 2 then $\underline{A}$ is dual.

A logic formula gives us a means of satisfying ourselves of the truth of number theoretic theorems. For to each number theoretic proposition there corresponds a W.F.F. $\underline{A}$ which is dual if and only if the proposition is true. Now if $\underline{L}$ is a logic and $\underline{L}(\underline{A})$ conv 2 then $\underline{A}$ is dual and we know that the corresponding number theoretic proposition is true. It does not follow that if $\underline{L}$ is a logic we can use $\underline{L}$ to satisfy ourselves of the truth of any true number theoretic theorem.

If $\underline{L}$ is a logic the set of formulae $\underline{A}$ for which $\underline{L}(\underline{A})$ conv 2 will be called the extent of $\underline{L}$.

It may be proved by the use of (D), (E) p 7, that there is a formula X such that if $\underline{M}$ has a normal form and no free variables and is not convertible to 2, then $X(\underline{M})$ conv 1, but if $\underline{M}$ conv 2 then $X(\underline{M})$ conv 2. If $\underline{L}$ is a logic then $\lambda x. X(\underline{L}(x))$ is also a logic, whose extent is the same as that of $\underline{L}$, and has the property that if $\underline{A}$ has no free variables then $\{\lambda x. X(\underline{L}(x))\}(\underline{A})$ is always convertible to 1 or to 2 or else has no normal form. A logic with this property will be said to be standardized.

We shall say that a logic $\underline{L}'$ is at least as complete as a logic $\underline{L}$ if the extent of $\underline{L}$ is a subset of the extent of $\underline{L}'$. The logic $\underline{L}'$ will be more complete than $\underline{L}$ if the extent of $\underline{L}$ is a proper subset of the extent of $\underline{L}'$.

Suppose that we have an effective set of rules by which we can prove formulae to be dual; i.e. we have a system of symbolic logic in which the propositions proved are of the form that certain formulae are dual. Then we can find a logic formula whose extent consists of just those formulae which can be proved to be dual by the rules; that is to say that there is a rule for obtaining the logic formula from the system of symbolic logic. In fact the system of symbolic logic enables us to obtain[10] a computable function of positive integers whose values run through the Gödel representations of the formulae provable by means of the given rules. By the theorem of equivalence of computable and λ-definable functions there is a formula $\underline{J}$ such that $\underline{J}(1), \underline{J}(2), \ldots$ are the G. R. of these formulae, Now let

$$W \rightarrow \lambda j v.\, \mathcal{P}(\lambda u.\, \delta(j(u), v), 1, \underline{I}, 2)$$

then I assert that $W(\underline{J})$ is a logic with the required properties. The properties of $\mathcal{P}$ imply that $\mathcal{P}(\underline{C}, 1)$ is convertible to the least positive integer $\underline{n}$ for which $\underline{C}(\underline{n})$ conv 2 and has no normal form if there is no such integer. Consequently $\mathcal{P}(\underline{C}, 1, \underline{I}, 2)$ is convertible to 2 if $\underline{C}(\underline{n})$ conv 2 for some positive integer n, and has no normal form otherwise. That is to say that $W(\underline{J}, \underline{A})$ conv 2 if and only if $\delta(\underline{J}(\underline{n}), \underline{A})$ conv 2, some n, i.e. if $\underline{J}(\underline{n})$ conv $\underline{A}$ some n.

There is conversely a formula W' such that if $\underline{L}$ is a logic

---

[12] Compare Turing [1], 252, second footnote, [2], 156.

then $W'(\underline{L})$ enumerates the extent of $\underline{L}$. For there is a formula Φ such that $\Phi(\underline{L}, \underline{A}, \underline{n})$ conv 2 if and only if $\underline{L}(\underline{A})$ is convertible to 2 in less than n steps. We then put

$$W' \rightarrow \lambda l n . \mathrm{form}(\mathfrak{D}(2, \mathfrak{B}(\lambda x . \Phi(l, \mathrm{form}(\mathfrak{D}(2,x)), \mathfrak{D}(3,x)), n))),$$

Of course $W'(W(\underline{J}))$ will normally be entirely different from $\underline{J}$ and $W(W'(\underline{L}))$ from $\underline{L}$.

In the case where we have symbolic logic whose propositions can be interpreted as number theoretic theorems, but are not expressed in the form of the duality of formulae we shall again have a corresponding logic formula, but its relation to the symbolic logic will not be so simple. As an example let us take the case that the symbolic logic proves that certain primitive recursive functions vanish infinitely often. As was shown in § 3 we can associate with each such proposition a W.F.F. which is dual if and only if the proposition is true. When we replace the propositions of the symbolic logic by theorems on the duality of formulae in this way our previous argument applies, and we obtain a certain logic formula $\underline{L}$. However, $\underline{L}$ does not determine uniquely which are the propositions provable in the symbolic logic; for it is possible that '$\theta_1(x)$ vanishes infinitely often' and '$\theta_2(x)$ vanishes infinitely often' are both associated with '$\underline{A}$ is dual', and that the first of these propositions is provable in the system, but the second not. However, if we suppose that the system of symbolic logic is sufficiently powerful to be able to carry out the argument on p. 15 then this difficulty cannot arise. There is also the possibility that there may be formulae in the extent of $\underline{L}$ with no propositions of the form '$\theta(x)$ vanishes infinitely often' corresponding to them. But to each such formula we can assign (by a different argument) a proposition P of the symbolic logic which is the necessary and sufficient condition for $\underline{A}$ to be dual. With P is associated (in the first way) a formula $\underline{A}'$. Now $\underline{L}$ can always be modified so that its extent contains $\underline{A}'$ whenever it contains $\underline{A}$.

We shall be interested principally in questions of completeness. Let us suppose that we have a class of systems of symbolic logic the propositions of these systems being expressed in a uniform notation and interpretable as number theoretic theorems; suppose also there is a rule by which we can assign to each proposition P of the notation a W.F.F. $\underline{A}_P$ which is dual if and only if P is true, and that to each W.F.F. $\underline{A}$ we can assign a proposition $P_{\underline{A}}$ which is the necessary and sufficient condition for $\underline{A}$ to be dual. $P_{\underline{A}_P}$ is to be expected to differ from P. To each symbolic logic C we can assign two logic formulae $\underline{L}_C$ and $\underline{L}_C'$. A formula $\underline{A}$ belongs to the extent of $\underline{L}_C$ if $P_{\underline{A}}$ is provable in C, while the extent of $\underline{L}_C'$ consists of all $\underline{A}_P$ where P is provable in C. Let us say that the class of symbolic logics is complete if each true proposition is provable in one of them: let us also say that a class of logic formulae is complete if the set theoretic sum of the extents of these logics includes all dual formulae. I assert that a necessary condition for a class of symbolic logics C to be complete is that the class of logics $\underline{L}_C$ be complete, while a sufficient condition is that the class of logics $\underline{L}'_C$ be complete. Let us suppose that the class of symbolic logics is complete; consider $P_{\underline{A}}$ where $\underline{A}$ is arbitrary but dual. It must be provable in one of the systems, C say. $\underline{A}$ therefore belongs to the extent of $\underline{L}_C$, ie. the class of logics $\underline{L}_C$ is complete. Now suppose the class of logics $\underline{L}'_C$ is complete. Let P be an arbitrary true proposition of the notation; $\underline{A}_P$ must belong to the extent of some $\underline{L}'_C$, and this means that P is provable in C.

We shall say that a single logic formula $\underline{L}$ is complete if its extent includes all dual formulae; that is to say that it is ~~dual~~ complete if it enables us to prove every true number theoretic theorem. It is a consequence of the theorem of Gödel (if suitably extended) that no logic formula is complete, and this also follows from (C) p. 6, or from the results of Turing [1] § 8, when taken in conjunction with § 3 of the present paper. The idea of completeness of a logic formula will not therefore be very important, although it is useful to have a term for it.

Suppose $\underline{Y}$ is a W.F.F. such that $\underline{Y}(\underline{n})$ is a logic for each positive integer $\underline{n}$. The formulae of the extent of $\underline{Y}(\underline{n})$ are enumerated by $W(\underline{Y}(\underline{n}))$, and the combined extents of these logics by $\lambda r.\, W(\underline{Y}(\mathfrak{D}(2,r)), \mathfrak{D}(3,r))$. Putting

$$\Gamma \rightarrow \lambda y.\, W'(\lambda r.\, W(y(\mathfrak{D}(2,r)), \mathfrak{D}(3,r)))$$

$\Gamma(\underline{Y})$ is a logic whose extent is the combined extent of $\underline{Y}(1)$, $\underline{Y}(2)$, $\underline{Y}(3)$, . . .

To each W.F.F. $\underline{L}$ we can assign a W.F.F. $V(\underline{L})$ such that the necessary and sufficient condition for $\underline{L}$ to be a logic formula is that $V(\underline{L})$ be dual. Let Nm be a W.F.F. which enumerates all formulae with normal forms. Then the condition that $\underline{L}$ be a logic is that $\underline{L}(Nm(\underline{r}), \underline{s})$ conv 2 for all positive integers r, s, i.e. that $\lambda a . \underline{L}(Nm(\mathfrak{D}(2,a)), \mathfrak{D}(3,a))$ be dual. We may therefore put

$$V \rightarrow \lambda la . l(Nm(\mathfrak{D}(2,a)), \mathfrak{D}(3,a))$$

7. Ordinals.

We begin our treatment of ordinals with some brief definitions from the Cantor theory of ordinals, but for the understanding of some of the proofs a greater amount of the Cantor theory is necessary than is here set out.

Suppose we have a class determined by the propositional function $D(x)$ and a relation $G(x,y)$ ordering them, i.e. satisfying

$$\left.\begin{array}{ll} G(x,y) \;\&\; G(y,z) \supset G(x,z) & \text{i} \\ D(x) \;\&\; D(y) \supset G(x,y) \vee G(y,x) \vee x=y & \text{ii} \\ G(x,y) \supset D(x) \;\&\; D(y) & \text{iii} \\ \sim G(x,x) & \text{iv} \end{array}\right\} \qquad (7.1)$$

The class defined by $D(x)$ is then called a <u>series</u> with the ordering relation $G(x,y)$. The series is said to be <u>well ordered</u> and the ordering relation is called an <u>ordinal</u> if every sub-series which is not void has a first term, i.e. if

$$(D')\{(\exists x)(D'(x)) \;\&\; (x)(D'(x) \supset D(x)) \supset$$
$$\supset (\exists z)(y)[D'(z) \;\&\; (D'(x) \supset G(z,y) \vee z=y)]\} \qquad (7.2)$$

The condition (7.2) is equivalent to another, more suitable for our purposes, namely the condition that every descending subsequence must terminate; formally

$$(x)\{D'(x) \supset D(x) \;\&\; (\exists y)(D'(y) \;\&\; G(y,x))\} \supset (x)(\sim D'(x)) \qquad (7.3)$$

The ordering relation $G(x,y)$ is said to be similar to $G'(x,y)$ if there is a one-one correspondence between the series transforming the one relation into the other. This is best expressed formally

$$(\exists M)\Big[(x)\{D(x)\supset(\exists x')M(x,x')\}\ \&\ (x')\{D'(x')\supset(\exists x)M(x,x')\}$$
$$\&\ \Big\{\big(M(x,x')\ \&\ M(x,x'')\big)\vee\big(M(x',x)\ \&\ M(x'',x)\big)\Big\}\supset x'=x''\Big\} \qquad (7.4)$$
$$\&\ \{M(x,x')\ \&\ M(y,y')\supset(G(x,y)\equiv G'(x',y'))\}\Big]$$

Ordering relations are regarded as belonging to the same ordinal if and only if they are similar.

We wish to give names to all the ordinals, but this will not be possible until they have been restricted in some way; the class of ordinals as at present defined is more than enumerable. The restrictions we actually put are these: $D(x)$ is to imply that x is a positive integer; $D(x)$ and $G(x,y)$ are to be computable properties. Both of the propositional functions $D(x)$, $G(x,y)$ can then be described by means of a single W.F.F. $\underline{\Omega}$ with the properties.

$\underline{\Omega}(\underline{m},\underline{n})$ conv 4 unless both $D(m)$ and $D(n)$ are true,

$\underline{\Omega}(\underline{m},\underline{m})$ conv 3 if $D(m)$ is true,

$\underline{\Omega}(\underline{m},\underline{n})$ conv 2 if $D(m)$, $D(n)$, $G(m,n)$, $\sim(m=n)$ are true,

$\underline{\Omega}(\underline{m},\underline{n})$ conv 1 if $D(m)$, $D(n)$, $\sim G(m,n)$, $\sim(m=n)$, are true,

Owing to the conditions to which $D(x)$, $G(x,y)$ are subjected $\underline{\Omega}$ must further satisfy

(a) if $\underline{\Omega}(\underline{m},\underline{n})$ is convertible to 1 or 2 then $\underline{\Omega}(\underline{m},\underline{m})$ and $\underline{\Omega}(\underline{n},\underline{n})$ are convertible to 3,

(b) if $\underline{\Omega}(\underline{m},\underline{m})$ and $\underline{\Omega}(\underline{n},\underline{n})$ are convertible to 3 then $\underline{\Omega}(\underline{m},\underline{n})$ is convertible to 1, 2, or 3,

(c) if $\underline{\Omega}(\underline{m},\underline{n})$ is convertible to 1 then $\underline{\Omega}(\underline{n},\underline{m})$ is convertible to 2 and conversely,

(d) if $\underline{\Omega}(\underline{m},\underline{n})$ and $\underline{\Omega}(\underline{n},\underline{p})$ are convertible to 1 then $\underline{\Omega}(\underline{m},\underline{p})$

is also,

(e) there is no sequence $m_1, m_2, \ldots$ such that $\underline{\Omega}(\underline{m}_{i+1}, \underline{m}_i)$ conv 2 for each positive integer i,

(f) $\underline{\Omega}(\underline{m}, \underline{n})$ is always convertible to 1, 2, 3, or 4.

If a formula $\underline{\Omega}$ satisfies these conditions then there are corresponding propositional functions $D(x)$, $G(x, y)$. We shall therefore say that $\underline{\Omega}$ is an <u>ordinal formula</u> if it satisfies the conditions (a) - (f). It will be seen that a consequence of this definition is that Dt is an ordinal formula. It represents the ordinal ω. The definition we have given does not pretend to have virtues such as elegance or convenience. It has been introduced rather to fix our ideas and to show how it is possible in principle to describe ordinals by means of well formed formulae. The definitions could be modified in a number of ways. Some such modifications are quite trivial; they are typified by modifications such as changing the numbers 1,2,3,4, used in the definition to some others. Two such definitions will be said to be equivalent; in general we shall say that two definitions are equivalent if there are W.F.F. $\underline{T}$, $\underline{T}'$ such that if $\underline{A}$ is an ordinal formula under one definition and represents the ordinal α, then $\underline{T}'(\underline{A})$ is an ordinal formula under the second definition and represents the same ordinal, and conversely if $\underline{A}'$ is an ordinal formula under the second definition representing α, then $\underline{T}(\underline{A}')$ represents α under the first definition. Besides definitions equivalent in this sense to our original definition there are a number of other possibilities open.

Suppose for instance that we do not require $\mathfrak{D}(x)$ and $G(x,y)$ to be computable, but only that $\mathfrak{D}(x)$ and $G(x,y) \& x < y$ be axiomatic.[13] This leads to a definition of ordinal formula which is (presumably) not equivalent to the definition we are using.[14] There are numerous possibilities, and little to guide us as to which definition should be chosen. No one of them could well be described as 'wrong'; some of them may be found more valuable in applications than others, and the particular choice we have made has been partly determined by the applications we have in view. In the case of theorems of a negative character one would wish to prove them for each one of the possible definitions of 'ordinal formula'. This program could I think be carried through for the negative results of § 9, 10.

[13] To require $G(x,y)$ to be axiomatic would amount to requiring $G(x,y)$ computable on account of (7.1) ii .

[14] On the other hand if $\mathfrak{D}(x)$ be axiomatic and $G(x,y)$ computable in the modified sense that there is a rule for determining whether $G(x,y)$ is true which leads to a definite result in all cases where $\mathfrak{D}(x)$ and $\mathfrak{D}(y)$ are true, the corresponding definition of ordinal formula is equivalent to our definition. To give the proof would be too much of a digression. Probably a number of other equivalences of this kind hold.

Before leaving the subject of possible ways of defining ordinal formulae I must mention another definition due to Church and Kleene (Church and Kleene [1]). We can make use of this definition in constructing ordinal logics, but it is more convenient to use a slightly different definition which is equivalent (in the sense described on p. 29) to the Church-Kleene definition as modified in Church [4].

Introduce the abbreviations

$$U \rightarrow \lambda u f x . u(\lambda y . f(y(I, x)))$$

$$Suc \rightarrow \lambda a u f x . f(a(u, f, x))$$

We define first a partial ordering relation '$<$' which holds between certain pairs of W.F.F. (conditions (1) - (5)).

(1) If $\underline{A}$ conv $\underline{B}$ then $\underline{A} < \underline{C}$ implies $\underline{B} < \underline{C}$ and $\underline{C} < \underline{A}$ implies $\underline{C} < \underline{B}$.

(2) $\underline{A} < Suc(\underline{A})$

(3) For any positive integers $m, n,$ $\lambda u f x . \underline{R}(\underline{n}) < \lambda u f x . \underline{R}(\underline{n}$ implies $\lambda u f x . \underline{B}(\underline{n}) < \lambda u f x . u(\underline{R})$.

(4) If $\underline{A} < \underline{B}$ and $\underline{B} < \underline{C}$ then $\underline{A} < \underline{C}$. (1) - (4) are required for any W.F.F. $\underline{A}, \underline{B}, C, \lambda u f x . \underline{R}$.

(5) The relation $\underline{A} < \underline{B}$ holds only when compelled to do so by (1) - (4).

We define C-K ordinal formulae by the conditions (6) - (10).

(6) If $\underline{A}$ conv $\underline{B}$ and $\underline{A}$ is a C-K ordinal formula then $\underline{B}$ is a C-K ordinal formula.

(7) U is a C-K ordinal formula.

(8) If $\underline{A}$ is a C-K ordinal formula then $Suc(\underline{A})$ is a C-K ordinal formula.

(9) If $\lambda u f x . \underline{R}(\underline{n})$ is a C-K ordinal formula and $\lambda u f x . \underline{R}(\underline{n}) < \lambda u f x . \underline{R}(S(\underline{n}))$ for each positive integer n then $\lambda u f x . u(\underline{R})$ is a C-K ordinal formula.

(10) A formula is a C-K ordinal formula only if compelled to be so by (6) - (9).

The representation of ordinals by formulae is described by (11) - (15).

(11) If $\underline{A}$ conv $\underline{B}$ and $\underline{A}$ represents α then $\underline{B}$ represents α.

(12) U represents 1.

(13) If $\underline{A}$ represents α then $Suc(\underline{A})$ represents $\alpha + 1$.

(14) If $\lambda u f x . \underline{B}(\underline{n})$ represents α_n for each positive integer n then $\lambda u f x . u(\underline{B})$, represents the upper bound of the sequence $\alpha_1, \alpha_2, \alpha_3 \ldots$.

(15) A formula represents an ordinal only when compelled to do so by (11) - (14).

We denote any ordinal represented by $\underline{A}$ by $\Xi_{\underline{A}}$ without prejudice to the possibility that more than one ordinal may be represented by $\underline{A}$. We shall write $\underline{A} \leq \underline{B}$ to mean $\underline{A} < \underline{B}$ or $\underline{A}$ conv $\underline{B}$.

In proving properties of C-K ordinal formulae we shall often use a kind of analogue of the principle of transfinite induction. If φ is some property and we have

(a) If $\underline{A}$ conv $\underline{B}$ and $\varphi(\underline{A})$ then $\varphi(\underline{B})$.

(b) $\varphi(U)$.

(c) If $\varphi(\underline{A})$ then $\varphi(Suc(\underline{A}))$.

(d) If $\varphi(\lambda u f x . \underline{B}(\underline{n}))$ and $\lambda u f x . \underline{B}(\underline{n}) < \lambda u f x . \underline{B}(S(\underline{n}))$ for each positive integer n then $\varphi(\lambda u f x . u(\underline{B}))$

then $\varphi(\underline{A})$ for each C-K ordinal formula $\underline{A}$. To prove the validity of this principle we have only to observe that the class of formulae $\underline{A}$ satisfying $\varphi(\underline{A})$ is one of those of which the class of C-K ordinal formulae was defined to be the smallest. We can use this principle to help us prove:-

(i) Every C-K ordinal formula is convertible to the form $\lambda ufx.\underline{B}$ where $\underline{B}$ is in normal form.

(ii) There is a method by which one can determine of any C-K ordinal formula into which of the forms U, $Suc(\lambda ufx.\underline{B})$, $\lambda ufx.u(\underline{R})$ where u is free in $\underline{R}$, it is convertible, and to determine $\underline{B}$, $\underline{R}$. In each case $\underline{B}$, $\underline{R}$ are unique apart from conversions.

(iii) If $\underline{A}$ represents any ordinal $\Xi_{\underline{A}}$ is unique. If $\Xi_{\underline{A}}$, $\Xi_{\underline{B}}$ exist and $\underline{A} < \underline{B}$ then $\Xi_{\underline{A}} < \Xi_{\underline{B}}$.

(iv) If $\underline{A}$, $\underline{B}$, $\underline{C}$ are C-K ordinal formulae and $\underline{B} < \underline{A}$, $\underline{C} < \underline{A}$ then either $< \underline{B}$, $\underline{B} < \underline{C}$ or $\underline{B}$ conv $\underline{C}$.

(v) A formula $\underline{A}$ is a C-K ordinal formula if

(A) $U \leq \underline{A}$

(B) If $\lambda ufx.u(\underline{R}) \leq \underline{A}$ and n is a positive integer, then $\lambda ufx.\underline{R}(\underline{n}) < \lambda ufx.\underline{R}(S(\underline{n}))$.

(C) For any two W.F.F. $\underline{B}$, $\underline{C}$ with $\underline{B} < \underline{A}$, $\underline{C} < \underline{A}$ we have $\underline{B} < \underline{C}$, $\underline{C} < \underline{B}$ or $\underline{B}$ conv $\underline{C}$, but never $\underline{B} < \underline{B}$.

(D) There is no infinite sequence $\underline{B}_1$, $\underline{B}_2$, . . . for which $\underline{B}_r < \underline{B}_{r-1} < \underline{A}$ each r.

(vi) There is a formula H such that if $\underline{A}$ is a C-K ordinal formula then $H(\underline{A})$ is an ordinal formula representing the same ordinal. $H(\underline{A})$ is not an ordinal formula unless $\underline{A}$ is a C-K ordinal formula.

<u>Proof of (i)</u>. Take $\varphi(\underline{A})$ to be ' $\underline{A}$ is convertible to the form

$\lambda ufx . \underline{B}$ where $\underline{B}$ is in normal form'. The conditions (a), (b) are trivial. For (c) suppose $\underline{A}$ conv $\lambda ufx . \underline{B}$ where $\underline{B}$ is in normal form, then $Suc(\underline{A})$ conv $\lambda ufx . f(\underline{B})$ and $f(\underline{B})$ is in normal form. For (d) we have only to show that $u(\underline{R})$ has a normal form, i.e. that $\underline{R}$ has a normal form, which is true since $\underline{R}(I)$ has a normal form.

Proof of (ii). Since by hypothesis the formula is a C-K ordinal formula we have only to perform conversions on it until it is in one of the forms described. It is not possible to convert it into two of these three forms. For suppose $\lambda ufx . f(\underline{A}(u, f, x))$ conv $\lambda ufx . u(\underline{R})$ and is a C-K ordinal formula; it is therefore (then) convertible to the form $\lambda ufx . \underline{B}$ where $\underline{B}$ is in normal form. But the normal form of $\lambda ufx . u(\underline{R})$ can be obtained by conversions on $\underline{R}$, and that of $\lambda ufx . f(\underline{A}(u, f, x))$ by conversions on $\underline{A}(u, f, x)$ (as follows from Church and Rosser [1] theorem 2) but this would imply that the formula in question had two normal forms, one of form $\lambda ufx . u(\underline{S})$ and one of form $\lambda ufx . f(\underline{C})$, which is impossible. Or suppose $\underline{U}$ conv $\lambda ufx . u(\underline{R})$ where $\underline{R}$ is a well formed formula with u as a free variable. We may suppose $\underline{R}$ is in normal form. Now $\underline{U}$ is $\lambda ufx . u(\lambda y . f(y(I, x)))$. By (A) p. 6. $\underline{R}$ is identical with $\lambda y . f(y(I, x))$ which does not have u as a free variable. It now only remains to show that if $Suc(\lambda ufx . \underline{B})$ conv $Suc(\lambda ufx . \underline{B}')$ and $\lambda ufx . u(\underline{R})$ conv $\lambda ufx . u(\underline{R}')$ then $\underline{B}$ conv $\underline{B}'$ and $\underline{R}$ conv $\underline{R}'$.

If $Suc(\lambda ufx . \underline{B})$ conv $Suc(\lambda ufx . \underline{B}')$ then $\lambda ufx . f(\underline{B})$ conv $\lambda ufx . f(\underline{B}')$

but both of these formulae can be brought to normal form by conversions on $\underline{B}$, $\underline{B}'$ and therefore $\underline{B}$ conv $\underline{B}'$. The same argument applies in the case that $\lambda u f x . u(\underline{R})$ conv $\lambda u f x . u(\underline{R}')$.

Proof of (iii). To prove the first half take $\varphi(\underline{A})$ to be '$\Xi_{\underline{A}}$ is unique'. (7.5) (a) is trivial and (b) follows from the fact that $\underline{U}$ is not convertible either to the form $Suc(\underline{A})$ or to $\lambda u f x . u(\underline{R})$ where $\underline{R}$ has u as a free variable. For (c): $Suc(\underline{A})$ is not convertible to the form $\lambda u f x . u(\underline{R})$; the possibility of $Suc(\underline{A})$ representing an ordinal on account of (12) or (14) is therefore eliminated. By (13) $Suc(\underline{A})$ represents $\alpha' + 1$ if $\underline{A}'$ represents α' and $Suc(\underline{A})$ conv $Suc(\underline{A}')$. If we suppose $\underline{A}$ represents α , then $\underline{A}$, $\underline{A}'$ being C-K ordinal formulae are convertible to the forms $\lambda u f x . \underline{B}'$, $\lambda u f x . \underline{B}'$ but then by (ii) $\underline{B}$ conv $\underline{B}'$ i.e. $\underline{A}$ conv $\underline{A}'$, and therefore by the hypothesis $\varphi(\underline{A})$, $\alpha = \alpha'$. Then $\Xi_{Suc(\underline{A})} = \alpha' + 1$ is unique. For (d): $\lambda u f x . u(\underline{R})$ is not convertible to the form $Suc(\underline{A})$ or to $\underline{U}$ if $\underline{R}$ has u as a free variable. If $\lambda u f x . u(\underline{R})$ represents an ordinal it is therefore in virtue of (14), possibly together with (11). Now if $\lambda u f x . u(\underline{R})$ conv $\lambda u f x . u(\underline{R}')$ then $\underline{R}$ conv $\underline{R}'$, so that the sequence $\lambda u f x . \underline{R}(1)$, $\lambda u f x . \underline{R}(2)$, . . . in (14) is unique apart from conversions. Then by the induction hypothesis the sequence α_1 , α_2, α_3 , . . . is unique. The only ordinal that is represented by $\lambda u f x . u(\underline{R})$ is the upper bound of this sequence which is unique.

For the second half we use a type of argument rather different from our transfinite induction principle. The formulae $\underline{B}$ for which $\underline{A} < \underline{B}$ form the smallest class for which

$Suc(\underline{A})$ belongs to the class.

If $\underline{C}$ belongs to the class then $Suc(\underline{C})$ belongs to it.

If $\lambda u f x . \underline{R}(\underline{n})$ belongs to the class and $\lambda u f x . \underline{R}(\underline{n}) < \lambda u f x . \underline{R}(\underline{m})$ where m, n are some positive integers then $\lambda u f x . u(\underline{R})$ belongs to it.

If $\underline{C}$ belongs to the class and $\underline{C}$ conv $\underline{C}'$ then $\underline{C}'$ belongs to it.

It will suffice to prove that the class of formulae $\underline{B}$ for which either $\Xi_{\underline{B}}$ does not exist or $\Xi_{\underline{A}} < \Xi_{\underline{B}}$ satisfies the conditions (7.6). Now

$$\Xi_{Suc(\underline{A})} = \Xi_{\underline{A}} + 1 > \Xi_{\underline{A}}$$

$$\Xi_{Suc(\underline{C})} > \Xi_{\underline{C}} > \Xi_{\underline{A}} \quad \text{if } \underline{C} \text{ is in the class.}$$

If $\Xi_{\lambda u f x . \underline{R}(\underline{n})}$ does not exist then $\Xi_{\lambda u f x . u(\underline{R})}$ does not exist, and therefore $\lambda u f x . u(\underline{R})$ is in the class. If $\Xi_{\lambda u f x . \underline{R}(\underline{n})}$ exists and is greater than $\Xi_{\underline{A}}$ and $\lambda u f x . \underline{R}(\underline{n}) < \lambda u f x . \underline{R}(\underline{m})$ then

$$\Xi_{\lambda u f x . u(\underline{R})} \geqslant \Xi_{\lambda u f x . \underline{R}(\underline{n})} > \Xi_{\underline{A}}$$

so that $\lambda u f x . u(\underline{R})$ belongs to the class.

Proof of (iv). We prove this by induction with respect to $\underline{A}$. Take $\varphi(\underline{A})$ to be 'whenever $\underline{B} < \underline{A}$ and $\underline{C} < \underline{A}$ then $\underline{B} < \underline{C}$ or $\underline{C} < \underline{B}$ or $\underline{B}$ conv $\underline{C}$'. $\varphi(U)$ follows from the fact that we never have $\underline{B} < U$. If we have $\varphi(\underline{A})$ and $\underline{B} < Suc(\underline{A})$ then either $\underline{B} < \underline{A}$ or $\underline{B}$ conv $\underline{A}$; for we can find $\underline{D}$ so that $\underline{B} \leq \underline{D}$, and $\underline{D} < Suc(\underline{A})$ can be proved without appealing either to (1) or (5); (4) does not apply so we must have $\underline{D}$ conv $\underline{A}$. Then if $\underline{B} < Suc(\underline{A})$ and $\underline{C} < Suc(\underline{A})$ we have four possibilities

$\underline{B}$ conv $\underline{A}$, $\underline{C}$ conv $\underline{A}$

$\underline{B}$ conv $\underline{A}$, $\underline{C} < \underline{A}$

$\underline{B} < \underline{A}$, $\underline{C}$ conv $\underline{A}$

$\underline{B} < \underline{A}$, $\underline{C} < \underline{A}$

In the first case $\underline{B}$ conv $\underline{C}$, in the second $\underline{C} < \underline{B}$, in the third $\underline{B} < \underline{C}$ and in the fourth the induction hypothesis applies.

Now suppose that $\lambda ufx.\underline{R}(\underline{n})$ is a C-K ordinal formula, $\lambda ufx.\underline{R}(\underline{n}) < \lambda ufx.\underline{R}(S(\underline{n}))$ and $\varphi(\underline{R}(\underline{n}))$, for each positive integer n, and $\underline{A}$ conv $\lambda ufx.u(\underline{R})$. Then if $\underline{B} < \underline{A}$ this means that $\underline{B} < \lambda ufx.\underline{R}(\underline{n})$ for some n; if we have also $\underline{C} < \underline{A}$ then $\underline{B} < \lambda ufx.\underline{R}(\underline{n}')$, $\underline{C} < \lambda ufx.\underline{R}(\underline{n}')$ some n'. Thus for these $\underline{B}$, $\underline{C}$ the required result follows from $\varphi(\lambda ufx.\underline{R}(\underline{n}'))$.

Proof of (v). The conditions (C), (D) imply that the classes of interconvertible formulae $\underline{B}$, $\underline{B} < \underline{A}$ are well-ordered by the relation '$<$'. We prove (v) by (ordinary) transfinite induction with respect to the order type α of the series formed by these classes; (α is in fact the solution of the equation $1 + \alpha = \Xi_{\underline{A}}$ but we do not need this). We suppose then that (v) is true for all order types less than α. If $\underline{E} < \underline{A}$ then $\underline{E}$ satisfies the conditions of (v) and the corresponding order type is smaller: $\underline{E}$ is therefore a C-K ordinal formula. This expresses all consequences of the induction hypothesis that we need. There are three cases to consider.

(x) $\alpha = 0$

(y) $\alpha = \beta + 1$

(z) α is of neither of the forms (x), (y).

In case (x) we must have $\underline{A}$ conv U on account of (A). In case (y) there is a formula $\underline{D}$ such that $\underline{D} < \underline{A}$, and $\underline{B} \leq \underline{D}$ whenever $\underline{B} < \underline{A}$. The relation $\underline{D} < \underline{A}$ must hold in virtue either of (1), (2), (3), or (4). It cannot be in virtue of (4) for then there would be $\underline{B}$, $\underline{B} < \underline{A}$, $\underline{D} < \underline{B}$ contrary to (C) taken in conjunction with the definition of $\underline{D}$. If it is in virtue of (3) then α is the upper bound of a sequence $\alpha_1, \alpha_2 \ldots$ of ordinals, which are increasing on account of (iii) and the conditions $\lambda u f x . \underline{R}(\underline{u}) < \lambda u f x . \underline{R}(S(\underline{u}))$ in (3). This is inconsistent with $\alpha = \beta + 1$. This means that (2) applies (after we have eliminated (1) by suitable conversions on $\underline{A}$, $\underline{D}$) and we see that $\underline{A}$ conv $Suc(\underline{D})$; but since $\underline{D} < \underline{A}$, $\underline{D}$ is a C-K ordinal formula, and $\underline{A}$ must therefore be a C-K ordinal formula by (B). Now take case (z). It is impossible that $\underline{A}$ be of form $Suc(\underline{D})$, for then we should have $\underline{B} < \underline{D}$ whenever $\underline{B} < \underline{A}$ which would mean that we had case (y). Since $U < \underline{A}$ there must be an $\underline{F}$ such that $\underline{F} < \underline{A}$ is demonstrable either by (2) or by (3) (after a possible conversion on $\underline{A}$); it must of course be demonstrable by (3). Then $\underline{A}$ is of form $\lambda u f x . u(\underline{R})$. By (3), (B) we see that $\lambda u f x . \underline{R}(\underline{n}) < \underline{A}$ for each positive integer n; each $\lambda u f x . \underline{R}(\underline{n})$ is therefore a C-K ordinal formula. Applying (9), (B) we see that $\underline{A}$ is a C-K ordinal formula.

Proof of vi. To prove the first half it suffices to find a method whereby from a C-K ordinal formula $\underline{A}$ we can find the corresponding ordinal formula $\underline{\Omega}$. For then there is a formula H_1 such that $H_1(\underline{a})$ conv $\underline{p}$ if a is the G.R. of $\underline{A}$ and $\underline{p}$ that of $\underline{\Omega}$. H is then to be defined by

$$H \rightarrow \lambda a.\ \mathrm{form}(H_1(Gr(a)))$$

The method for finding $\underline{\Omega}$ may be replaced by a method of finding $\underline{\Omega}(\underline{m}, \underline{n})$ given $\underline{A}$ and any two positive integers m , n . We shall arrange the method so that whenever $\underline{A}$ is not an ordinal formula either the calculation of the values does not comes to an end or else the values are not consistent with $\underline{\Omega}$ being an ordinal formula. In this way we can prove the second half of (vi).

Let Ls be a formula such that $Ls(\underline{A})$ enumerates the classes of formulae $\underline{B}$, $\underline{B} < \underline{A}$ (i.e. if $\underline{B} < \underline{A}$ there is one and only one positive integer n for which $Ls(\underline{A}, \underline{n})$ conv $\underline{B}$). Then the rule for finding the value of $\underline{\Omega}(\underline{m}, \underline{n})$ is as follows:-

First determine whether $U \leq \underline{A}$ and whether $\underline{A}$ is convertible to the form $\underline{r}(Suc, U)$. This comes to an end if $\underline{A}$ is a C-K ordinal formula.

If $\underline{A}$ conv $\underline{r}(Suc, U)$ and either $m > r+1$ or $n > r+1$ then the value is 4. If $m < n \leq r+1$ the value is 2. If $n < m \leq r$ the value is 1. If $m = n \leq r+1$ the value is 3.

If $\underline{A}$ is not convertible to this form we determine whether either $\underline{A}$ or $Ls(\underline{A}, \underline{m})$ is convertible to the form $\lambda ufx.\,u(\underline{R})$ and if either of them is we verify that $\lambda ufx.\,\underline{R}(\underline{n}) < \lambda ufx.\,\underline{R}(S(\underline{n}$ We shall eventually comes to an affirmative answer if $\underline{A}$ is a C-K ordinal formula.

Having checked this we determine of m, n whether $Ls(\underline{A}, \underline{m}) < Ls(\underline{A}, \underline{n})$ $Ls(\underline{A}, \underline{n}) < Ls(\underline{A}, \underline{m})$, or $m = n$, and the value is to be accordingly 1, 2, or 3.

If $\underline{A}$ is a C-K ordinal formula this process certainly comes to an end. To see that the values so calculated correspond to an ordinal formula, and one representing $\Xi_{\underline{A}}$, first observe that this is so when $\Xi_{\underline{A}}$ is finite. In the other case (iii), (iv) show that $\Xi_{\underline{B}}$ determines a one-one correspondence between the ordinals β, $1 \leq \beta \leq \Xi_{\underline{A}}$ and the classes of interconvertible formulae $\underline{B}$, $\underline{B} < \underline{A}$. If we take $G(m, n)$ to be $Ls(\underline{A}, \underline{m}) < Ls(\underline{A}, \underline{n})$ we see that $G(m, n)$ is the ordering relation of a series of order type[15] $\Xi_{\underline{A}}$ and on the other

[15] The order type is β where $1 + \beta = \Xi_{\underline{A}}$ but $\beta = \Xi_{\underline{A}}$ since $\Xi_{\underline{A}}$ is infinite.

hand that the values of $\underline{\Omega}(\underline{m}, \underline{n})$ are related to $G(m, n)$ as on p. 29.

To prove the second half suppose $\underline{A}$ is not a C-K ordinal formula. Then one of the conditions (A)-(D) in (v) must not be satisfied. If (A) is not satisfied we shall not obtain a result even in the calculation of $\underline{\Omega}(\underline{1}, \underline{1})$. If (B) is not satisfied, for some positive integers $\underline{p}$, $\underline{q}$ we shall have $Ls(\underline{A}, \underline{p})$ conv $\lambda uf x . u(\underline{R})$ but not $\lambda uf x . \underline{R}(\underline{q}) < \lambda uf x . \underline{R}(S(\underline{q}))$. Then the process of calculating $\underline{\Omega}(\underline{p}, \underline{q})$ will not come to an end. In case of failure of (C) or (D) the values of $\underline{\Omega}(\underline{m}, \underline{n})$ may all be calculable but condition (b), (d), or (e) p. 29, 30 will be violated. Thus if $\underline{A}$ is not a C-K ordinal formula then $H(\underline{A})$ is not an ordinal formula.

I propose now to define three formulae Sum, Lim, Inf of importance in connection with ordinal formulae. As they are comparatively simple they will for once be given almost in full:

The formula Ug is one with the property that $Ug(\underline{m})$ is convertible to the formula representing the largest odd integer dividing m: it is not given in full. P is the predecessor function, $P(S(\underline{m}))$ conv $\underline{m}$

$$Al \rightarrow \lambda pxy . p(\lambda guv . g(v,u), \lambda uv . u(I,v), x, y)$$

$$Hf \rightarrow \lambda m . P(m(\lambda guv . g(v, S(u)), \lambda uv . v(I,u), 1, 2))$$

$$Bd \rightarrow \lambda w w' a a' x . Al(\lambda f . w(a, a, w'(a', a', f)), x, 4)$$

$$Sum \rightarrow \lambda w w' pq . Bd(w, w', Hf(p), Hf(q), Al(p, Al(q, w'(Hf(p), Hf(q), 1), Al(q, 2, w(Hf(p), Hf(q))))))$$

$$Lim \rightarrow \lambda zpq . \{\lambda ab . Bd(z(a), z(b), Ug(p), Ug(q), Al(Dt(a,b) + Dt(b,a), Dt(a,b), z(a, Ug(p), Ug(q)))))\}(\varpi(2,p), \varpi(2,q))$$

$$Inf \rightarrow \lambda wapq . Al(\lambda f . w(a, p, w(a, q, f)), w(p,q), 4)$$

The essential properties of these formulae are described by

$Al(2\underline{r}-1, \underline{m}, \underline{n})$ conv $\underline{m}$ $\qquad$ $Al(2\underline{r}, \underline{m}, \underline{n})$ conv $\underline{n}$

$Hf(2\underline{m})$ conv $\underline{m}$ $\qquad$ $Hf(2\underline{m}-1)$ conv $\underline{m}$

$Bd(\underline{\Omega}, \underline{\Omega}', \underline{a}, \underline{a}', x)$ conv 4 unless both $\underline{\Omega}(\underline{a}, \underline{a})$ conv 3 and $\underline{\Omega}'(\underline{a}', \underline{a}')$ conv 3 in which case it is convertible to x.

If $\underline{\Omega}$, $\underline{\Omega}'$ are ordinal formulae representing α, β respectively then $Sum(\underline{\Omega}, \underline{\Omega}')$ is an ordinal formula representing $\alpha + \beta$. If $\underline{Z}$ is a W.F.F. enumerating a sequence of ordinal formulae representing $\alpha_1, \alpha_2, \ldots$, then $Lim(\underline{Z})$ is an ordinal formula representing the infinite sum $\alpha_1 + \alpha_2 + \alpha_3 + \ldots$. If $\underline{\Omega}$ is an ordinal formula representing α then $Inf(\underline{\Omega})$ enumerates a sequence of ordinal formulae representing all the ordinals less than α without repetitions.

To prove that there is no general method for determining of a formula whether it is an ordinal formula we use an argument akin to that leading to the Burali-Forti paradox, but the emphasis and the conclusion are different. Let us suppose that such an algorithm is available. This enables us to obtain a recursive enumeration $\underline{\Omega}_1, \underline{\Omega}_2, \ldots$ of the ordinal formulae in normal form. There is a formula $\underline{Z}$ such that $\underline{Z}(\underline{n})$ conv $\underline{\Omega}_n$. Now $Lim(\underline{Z})$ represents an ordinal greater than any represented by an $\underline{\Omega}_n$, and has therefore been omitted from the enumeration.

This argument proves more than was originally asserted. In fact it proves that if we take any class $\underline{E}$ of ordinal formulae in normal form, such that if $\underline{A}$ is any ordinal formula then there is a formula in $\underline{E}$ representing the same ordinal as $\underline{A}$, then there is no method whereby one can tell whether a W.F.F. in normal form belongs to $\underline{E}$.

8. Ordinal logics.

An ordinal logic is a W.F.F. $\underline{\Lambda}$ such that $\underline{\Lambda}(\underline{\Omega})$ is a logic formula whenever $\underline{\Omega}$ is an ordinal formula.

This definition is intended to bring under one heading a number of ways of constructing logics which have recently been proposed or are suggested by recent advances. In this section I propose to show how to obtain some of these ordinal logics.

Suppose we have a class W of logical systems. The symbols used in each of these systems are the same, and a class of sequences of symbols called 'formulae' is defined, independently of the particular system in W. The rules of procedure of a system C define an axiomatic subset of the formulae, they are to be described as the 'provable formulae of C '. Suppose further that we have a method whereby, from any system of C of W we can obtain a new system C', also in W, and such that the set of provable formulae of C' include the provable formulae of C (we shall be most interested in the case where they are included as a proper subset.) It is to be understood that this 'method' is an effective procedure for obtaining the rules of procedure of C' from those of C.

Suppose that to certain of the formulae of W we make correspond number theoretic theorems: by modifying the definition of formula we may suppose that this is done for all formulae. We shall say that one of the systems C is valid if the provability of a formula in C implies the truth of the corresponding number theoretic theorem. Now let the relation of C' to C be such that the validity of C implies the validity of C', and let there be a valid system C_0 in W. Finally suppose that given any computable sequence $C_1, C_2, \ldots$ of systems in W the 'limit system' in which a formula is provable if and only if it is provable in one of the systems C_j also belongs to W. These limit systems are to be regarded, not as functions of the sequence given in extension, but as functions of the rules of formation of their terms. A sequence given in extension may be described by various rules of formation, and there will be several corresponding limit systems. Each of these may be described as a limit system of the sequence.

Under these circumstances we may construct an ordinal logic. Let us associate positive integers with the systems, in such a way that to each C corresponds a positive integer m_C, and m_C completely describes the rules of procedure of C. Then there is a W.F.F. $\underline{K}$, such that $\underline{K}(\underline{m}_C)$ conv $\underline{m}_{C'}$ for each C in W, and there is a W.F.F. $\underline{\Theta}$ such that if $\underline{D}(\underline{r})$ conv $\underline{m}_{C_r}$ for each positive integer r then $\underline{\Theta}(\underline{D})$ conv $\underline{m}_C$ where C is a limit system of $C_1, C_2, \ldots$. With each system C of W it is possible to associate a logic formula $\underline{L}_C$: the relation between them is that if G is a formula of W and the number theoretic theorem corresponding to G (assumed expressed in the conversion calculus form) asserts that $\underline{B}$ is dual, then $\underline{L}_C(\underline{B})$ conv 2 if and only if G is provable in C. There will be a W.F.F. $\underline{G}$ such that $\underline{G}(\underline{m}_C)$ conv $\underline{L}_C$ for each C of W. Put

$$\underline{N} \rightarrow \lambda a . \underline{G}(a(\underline{\Theta}, \underline{K}, \underline{m}_{C_0}))$$

I assert that $\underline{N}(\underline{A})$ is a logic formula for each C-K ordinal formula $\underline{A}$, and that if $\underline{A} < \underline{B}$ then $N(\underline{B})$ is more complete than $N(\underline{A})$, provided that there are formulae provable in C' but not in C for each valid C of W.

To prove this we shall show that to each C-K ordinal formula there corresponds a unique system $C[\underline{A}]$ such that

(i) $\underline{A}(\underline{\textcircled{H}}, \underline{K}, \underline{m}_{C_0})$ conv $\underline{m}_{C_0'}$

and that it further satisfies

(ii) $C[U]$ is a limit system of C_0', C_0', . . .

(iii) $C[Suc(\underline{A})]$ is $(C[\underline{A}])'$

(iv) $C[\lambda u f x . u(\underline{R})]$ is a limit system of $C[\lambda u f x . \underline{R}(\underline{1})]$, $C[\lambda u f x . \underline{R}(\underline{2})]$, . . . ,

$\underline{A}$ and $\lambda u f x . u(\underline{R})$ being assumed to be C-K ordinal formulae. The uniqueness of the system follows from the fact that $\underline{m}_C$ determines C completely. Let us try to prove the existence of $C[\underline{A}]$ for each C-K ordinal formula $\underline{A}$. As we have seen (p.33) it suffices to prove

(a) $C[U]$ exists,

(b) if $C[\underline{A}]$ exists then $C[Suc(\underline{A})]$ exists,

(c) if $C[\lambda u f x . \underline{R}(\underline{1})]$, $C[\lambda u f x . \underline{R}(\underline{2})]$, . . . exist then $C[\lambda u f x . u(\underline{R})]$ exists.

<u>Proof of (a)</u>.

$$\{\lambda y . \underline{K}(y(\underline{I}, \underline{m}_{C_0}))\}(\underline{n}) \quad \text{conv} \quad \underline{K}(\underline{m}_{C_0}) \quad \text{conv} \quad \underline{m}_{C}$$

for all positive integers n, and therefore by the definition of $\textcircled{H}$ there is a system, which we will call $C[U]$, and which is a limit system of C_0', C_0', . . . , satisfying

$$\underline{\Theta}(\lambda y.\,\underline{K}(y(\underline{I},\underline{m}_{C_0}))) \text{ conv } \underline{m}_{C[U]}$$

But on the other hand

$$U(\underline{\Theta},\underline{K},\underline{m}_{C_0}) \text{ conv } \underline{\Theta}(\lambda y.\,\underline{K}(y(\underline{I},\underline{m}_{C_0})))$$

This proves (a) and incidentally (ii)

Proof of (b).

$$\begin{aligned} Suc(\underline{A},\underline{\Theta},\underline{K},\underline{m}_{C_0}) &\text{ conv } \underline{K}(\underline{A}(\underline{\Theta},\underline{K},\underline{m}_{C_0})) \\ &\text{ conv } \underline{K}(\underline{m}_{C[\underline{A}]}) \\ &\text{ conv } \underline{m}_{(C[\underline{A}])'} \end{aligned}$$

Hence $C[Suc(\underline{A})]$ exists and is given by (iii).

Proof of (c).

$$\begin{aligned} \{\{\lambda ufx.\,\underline{R}\}(\underline{\Theta},\underline{K}\,\underline{m}_{C_0})\}(\underline{u}) &\text{ conv } \{\lambda ufx.\,\underline{R}(\underline{u})\}(\underline{\Theta},\underline{K},\underline{m}_{C_0}) \\ &\text{ conv } \underline{m}_{C[\lambda ufx.\,\underline{R}(\underline{u})]} \end{aligned}$$

by hypothesis. Consequently by the definition of $\underline{\Theta}$ there exists C which is a limit system of $C[\lambda ufx.\,\underline{R}(1)], C[\lambda ufx.\,\underline{R}(2)], \ldots$ and satisfies

$$\underline{\Theta}(\{\lambda ufx.\,\underline{R}\}(\underline{\Theta},\underline{K},\underline{m}_{C_0})) \text{ conv } \underline{m}_C$$

We define $C[\lambda ufx.\,u(\underline{R})]$ to be this C. We then have (iv) and

$$\begin{aligned} \{\lambda ufx.\,u(\underline{R})\}(\underline{\Theta},\underline{K},\underline{m}_{C_0}) &\text{ conv } \underline{\Theta}(\{\lambda ufx.\,\underline{R}\}(\underline{\Theta},\underline{K},\underline{m}_{C_0})) \\ &\text{ conv } \underline{m}_{C[\lambda ufx.\,u(\underline{R})]} \end{aligned}$$

This completes the proof of the properties (i) - (iv). From (ii), (iii), (iv) the facts that C_0 is valid and that C' is valid when C is valid we infer that $C[\underline{A}]$ is valid for each C-K ordinal formula $\underline{A}$: also that there are more formulae provable in $C[\underline{B}]$ than in $C[\underline{A}]$ when $\underline{A} < \underline{B}$. The truth of our assertions regarding $\underline{N}$ follows now in view of (i) and the definitions of $\underline{N}$ and $\underline{G}$.

We cannot conclude that $\underline{N}$ is an ordinal logic, since the formulae $\underline{A}$ were C-K ordinal formulae, but the formula H enables us to obtain an ordinal logic from $\underline{N}$. By the use of the formula Gr we obtain a formula Tn such that if $\underline{A}$ has a normal form then $Tn(\underline{A})$ enumerates the G.Rs. of the formulae into which $\underline{A}$ is convertible. Also there is a formula Ck such that if h is the G.R. of a formula $H(\underline{B})$ then $Ck(\underline{h})$ conv $\underline{B}$, but otherwise $Ck(\underline{h})$ conv U. Since $H(\underline{B})$ is an ordinal formula only if $\underline{B}$ is a C-K ordinal formula, $Ck(Tn(\underline{\Omega}, \underline{n}))$ is a C-K ordinal formula for each ordinal formula $\underline{\Omega}$ and integer n. For many ordinal formulae it will be convertible to U, but for suitable $\underline{\Omega}$, n it will be convertible to any given C-K ordinal formula. If we put

$$\underline{\Lambda} \rightarrow \lambda w a . \Gamma(\lambda n . \underline{N}(Ck(Tn(w, n)), a))$$

$\underline{\Lambda}$ will be the required ordinal logic. In fact on account of the properties of Γ, $\underline{\Lambda}(\underline{\Omega}, \underline{A})$ will be convertible to 2 if and only if there is a positive integer n such that

$$\underline{N}(Ck(Tn(\underline{\Omega}, \underline{n})), \underline{A}) \text{ conv } 2$$

If $\underline{\Omega}$ conv $H(\underline{B})$ there will be an integer n such that

$Ck(Tn(\underline{\Omega}, \underline{n}))$ conv $\underline{B}$, and then

$\underline{N}(Ck(Tn(\underline{\Omega}, \underline{n})), \underline{A})$ conv $\underline{N}(\underline{B}, \underline{A})$

For any n, $Ck(Tn(\underline{\Omega}, \underline{n}))$ is convertible to U or to some $\underline{B}$ where $\underline{\Omega}$ conv $H(\underline{B})$. Thus $\underline{\Lambda}(\underline{\Omega}, \underline{A})$ conv 2 if $\underline{\Omega}$ conv $H(\underline{B})$ and $\underline{N}(\underline{B}, \underline{A})$ conv 2 or if $\underline{N}(U, \underline{A})$ conv 2, but not in any other case.

We may now specialize and consider particular classes W of systems. First let us try to construct the ordinal logic described roughly in the introduction. For W we take the class of systems arising from the system of Principia Mathematica[16] by adjoining to it axiomatic (in the sense described on p. 10) sets of axioms[17]. Gödel has shown that primitive recursive relations[18] can be expressed by means of formulae in P. In fact there is a rule whereby given the recursion equations defining a primitive recursive relation we can find a formula[19] $\mathfrak{A}[x_0, \ldots, z_0]$ such that $\mathfrak{A}[f^{(m_1)}0, \ldots, f^{(m_r)}0]$ is provable in P if $F(m_1, \ldots, m_r)$ is true, and its negation is provable otherwise. Further there is a method by which one can tell of a formula $\mathfrak{A}[x_0, \ldots, z_0]$ whether it arises from a primitive recursive relation in this way, and by which one can find the equations

[16] Whitehead and Russell [1]. The axioms and rules of procedure of a similar system P will be found in a convenient form in Gödel [1]. I shall follow Gödel. The symbols for the natural numbers in P are $0, f0, ff0, \ldots f^{(n)}0 \ldots$. Variables with the suffix 'o' stand for natural numbers.

[17] It is sometimes regarded as necessary that the set of axioms used be computable, the intention being that it should be possible to verify of a formula reputed to be an axiom whether it really is so. We can obtain the same effect with axiomatic sets of axioms in this way. In the rules of procedure describing which are the axioms we incorporate a method of enumerating them, and we also introduce a rule that in the main part of the deduction whenever we write down an axiom as such we must also write down its position in the enumeration. It is possible to verify whether this has been done correctly.

[18] A relation $F(m_1, \ldots, m_r)$ is primitive recursive if it is the necessary and sufficient condition for the vanishing of a primitive recursive function $\varphi(m_1, \ldots, m_r)$.

[19] Capital German letters will be used to stand for variable or undetermined formulae in P. An expression such as $\mathfrak{A}[\mathfrak{B}, \mathfrak{C}]$ will stand for the result of substituting $\mathfrak{B}$ and $\mathfrak{C}$ for x_0 and y_0 in $\mathfrak{A}$.

which defined the relation. Formulae of this kind will be called <u>recursion formulae</u>. We shall make use of a property they have, which we cannot prove formally here without giving their definition in full, but which is essentially trivial. $\mathfrak{Db}[x_0, y_0]$ is to stand for a certain recursion formula such that $\mathfrak{Db}[f^{(m)}0, f^{(n)}0]$ is provable in P if $m = 2n$ and its negation is provable otherwise. Suppose that $\mathfrak{A}[x_0]$, $\mathfrak{B}[x_0]$ are two recursion formulae. Then the theorem I am assuming is that there is a recursion relation $\mathfrak{L}_{\mathfrak{A},\mathfrak{B}}[x_0]$ such that we can prove

$$\mathfrak{L}_{\mathfrak{A},\mathfrak{B}}[x_0] \equiv (\exists y_0)\Big(\big(\mathfrak{Db}[x_0, y_0] . \mathfrak{A}[y_0]\big) \vee \big(\mathfrak{Db}[fx_0, fy_0] . \mathfrak{B}[y_0]\big)\Big) \tag{8.1}$$

in P.

The significant formulae in any of our extensions of P are those of the form

$$(x_0)(\exists y_0)\, \mathfrak{A}[x_0, y_0] \tag{8.2}$$

where $\mathfrak{A}[x_0, y_0]$ is a recursion formula, arising from the relation $R(m, n)$ let us say. The corresponding number theoretic theorem states that for each natural number m there is a natural number n such that $R(m, n)$ is true.

The systems in W which are not valid are those in which a formula of form (8.2) is provable, but at the same time there is a natural number, m say, such that for each natural number n, $R(m, n)$ is false. This means to say that $\sim \mathfrak{A}[f^{(m)}0, f^{(n)}0]$ is provable for each natural number n. Since (8.2) is provable

$(\exists y_0)\,\mathfrak{A}[f^{(m)}0, y_0]$ is provable, so that

$$(\exists y_0)\,\mathfrak{A}[f^{(m)}0, y_0], \sim\mathfrak{A}[f^{(m)}0, 0], \sim\mathfrak{A}[f^{(m)}0, f0], \ldots \quad (8.3)$$

are all provable in the system. We may simplify (8.3). For a given m we may prove a formula of form $\mathfrak{A}[f^{(m)}0, y_0] \equiv \mathfrak{B}[y_0]$ in P, where $\mathfrak{B}[y_0]$ is a recursion formula. Thus we find that the necessary and sufficient condition for a system of W to be valid is that for no recursion formula $\mathfrak{B}[x_0]$ are all of the formulae

$$(\exists x_0)\,\mathfrak{B}[x_0], \sim\mathfrak{B}[0], \sim\mathfrak{B}[f0], \ldots \quad (8.4)$$

provable. An important consequence of this is that if

$\mathfrak{A}_1[x_0],\ \mathfrak{A}_2[x_0], \ldots, \mathfrak{A}_n[x_0]$

are recursion formulae and

$$(\exists x_0)\,\mathfrak{A}_1[x_0] \vee (\exists x_0)\,\mathfrak{A}_2[x_0] \vee \ldots \vee (\exists x_0)\,\mathfrak{A}_n[x_0] \quad (8.5)$$

is provable in C, and C is valid, then we can prove $\mathfrak{A}_r[f^{(a)}0]$ in C for some natural numbers r, a where $1 \leq r \leq n$. Let us define $\mathfrak{D}_r$ to be the formula

$(\exists x_0)\,\mathfrak{A}_1[x_0] \vee \ldots \vee (\exists x_0)\,\mathfrak{A}_r[x_0]$

and define $\mathfrak{E}_r[x_0]$ recursively by the condition that $\mathfrak{E}_1[x_0]$ be $\mathfrak{A}_1[x_0]$ and $\mathfrak{E}_{r+1}[x_0]$ be $\mathfrak{L}_{\mathfrak{E}_r, \mathfrak{A}_{r+1}}[x_0]$. Now I say that

$$\mathfrak{D}_r \supset (\exists x_0)\,\mathfrak{E}_r[x_0] \quad (8.6)$$

is provable for $1 \leq r \leq n$. It is clearly provable for $r = 1$: suppose it provable for a given r. We can prove

$$(y_0)(\exists x_0)\ \mathrm{Db}[x_0, y_0]$$

and

$$(y_0)(\exists x_0)\, Db(f x_0, f y_0)$$

from which we obtain

$$\mathcal{E}_r[y_0] \supset (\exists x_0)((Db[x_0, y_0] \cdot \mathcal{E}_r[y_0]) \vee (Db[x_0, y_0] \cdot \mathfrak{A}_{r+1}[y_0],$$

and

$$\mathfrak{A}_{r+1}[y_0] \supset (\exists x_0)((Db[x_0, y_0] \cdot \mathcal{E}_r[y_0]) \vee (Db[x_0, y_0] \cdot \mathfrak{A}_{r+1}[y_0],$$

These together with (8.1) yield

$$(\exists y_0)\, \mathcal{E}_r[y_0] \vee (\exists y_0)\, \mathfrak{A}_{r+1}[y_0] \supset (\exists x_0)\, \alpha_{\mathcal{E}_r, \mathfrak{A}_{r+1}}[x_0$$

which suffices to prove (8.6) for $r+1$. Now since (8.5) is provable in C , $(\exists x_0)\, \mathcal{E}_n[x_0]$ must be also, and since C is valid this means that $\mathcal{E}_n[f^{(m)}0]$ must be provable for some natural number m . From (8.1) and the definition of $\mathcal{E}_n[x_0]$ we see that this implies that $\mathfrak{A}_r[f^{(a)}0]$ is provable for some natural number a , and integer r , $1 \leq r \leq n$.

To any system C of W we can assign a primitive recursive relation $P_C(m, n)$ with the intuitive meaning 'm is the G.R. of a proof of the formula whose G.R. is n '. The corresponding recursion formula is $\mathrm{Proof}_C[x_0, y_0]$ (i.e. $\mathrm{Proof}_C[f^{(m)}0, f^{(n)}0]$ is provable when $P_C(m, n)$ is true, and its negation is provable otherwise). We can now explain what is the relation of a system C' to its predecessor C . The set of axioms which we adjoin to P to obtain C' consists of those adjointed in obtaining C , together with all formulae of the form

$$(\exists x_0)\, \mathrm{Proof}_C[x_0, f^{(m)}0] \supset \mathcal{F} \qquad (8.7)$$

where m is the G.R. of $\mathcal{F}$.

We wish to show that a contradiction can be obtained by assuming C' to be invalid but C to be valid. Let us suppose that a set of formulae of form (8.4) is provable in C'. Let $\mathfrak{A}_1, \mathfrak{A}_2, \ldots \mathfrak{A}_k$ be those axioms of C' of form (8.7) which are used in the proof of $(\exists x_0)\mathfrak{B}[x_0]$. We may suppose that none of them are provable in C. Then by the deduction theorem we see that

$$(\mathfrak{A}_1 . \mathfrak{A}_2 \ldots . \mathfrak{A}_k) \supset (\exists x_0)\mathfrak{B}[x_0] \qquad (8.8)$$

is provable in C. Let $\mathfrak{A}_l$ be $(\exists x_0)\,\mathrm{Proof}_C[x_0, f^{(m_l)}0] \supset \mathfrak{F}_l$ Then from (8.8) we find that

$$(\exists x_0)\,\mathrm{Proof}_C[x_0, f^{(m_1)}0] \vee \ldots \vee (\exists x_0)\,\mathrm{Proof}_C[x_0, f^{(m_k)}0] \vee (\exists x_0)\mathfrak{B}[x_0]$$

is provable in C. It follows from a result we have just proved that either $\mathfrak{B}[f^{(c)}0]$ is provable for some natural number c, or else $\mathrm{Proof}_C[f^{(u)}0, f^{(m_l)}0]$ is provable in C for some natural number u and some l, $1 \leqslant l \leqslant k$: but this would mean that $\mathfrak{F}_l$ was provable in C (this is one of the points where we assume the validity of C) and therefore also in C', contrary to hypothesis. Then $\mathfrak{B}[f^{(c)}0]$ must be provable in C; but we are also assuming $\sim \mathfrak{B}[f^{(c)}0]$ is provable in C'. There is therefore a contradiction in C'. Let us suppose that the axioms $\mathfrak{A}'_1 \ldots \mathfrak{A}'_{k'}$ of form (8.7) when adjoined to C suffice to obtain the contradiction and that none of these axioms are provable in C. Then

$$\sim \mathfrak{A}'_1 \vee \sim \mathfrak{A}'_2 \vee \ldots \vee \sim \mathfrak{A}'_{k'}$$

is provable in C, and if $\mathfrak{A}'_l$ is $(\exists x_0)\,\mathrm{Proof}_C[x_0, f^{(m'_l)}0] \supset \mathfrak{F}'_l$ then

$$(\exists x_0)\,\mathrm{Proof}_C[x_0, f^{(m'_1)}0] \vee \ldots \vee (\exists x_0)\,\mathrm{Proof}[x_0, f^{(m'_{k'})}0]$$

is provable in C. But by repetition of a previous argument this means that $\mathfrak{A}'_l$ is provable for some l, $1 \leq l \leq k'$ contrary to hypothesis. This is the required contradiction.

We may now construct an ordinal logic in the manner described on p. 44-48. But let us carry out the construction in rather more detail, and with some modifications appropriate to the particular case. Each system C of our set W may be described by means of a W.F.F. M_C which enumerates the G.Rs. of the axioms of C. There is a W.F.F. E such that if a is the G.R. of some proposition $\mathfrak{F}$ then $E(M_C, \underline{a})$ is convertible to the G.R. of

$$(\exists x_0)\,\mathrm{Proof}_C[x_0, f^{(a)}0] \supset \mathfrak{F}$$

If a is not the G.R. of any proposition in P then $E(M_C, \underline{a})$ is to be convertible to the G.R. of 0 = 0. From E we obtain a W.F.F. K such that $K(M_C, 2\underline{n}+1)$ conv $M_C(\underline{n})$, $K(M_C, 2\underline{n})$ conv $E(M_C, \underline{n})$. The successor system C' is defined by $K(M_C)$ conv $M_{C'}$. Let us choose a formula G such that $G(M_C, \underline{A})$ conv 2 if and only if the number theoretic theorem equivalent to '$\underline{A}$ is dual' is provable in C. Then we define Λ_P by

$$\Lambda_P \to \lambda w a.\,\Gamma(\lambda y.\,G(Ck(Tn(w,y), \lambda m n.m(\varpi(2,n), \varpi(3,n)), K, M_P)))$$

This is an ordinal logic provided that P is valid.

Another ordinal logic of this type has in effect been introduced by Church[20]. Superficially this ordinal logic seems to have no more

[20] In outline Church [1], 279-280. In greater detail Church [2], Chap. X.

in common with Λ_P than that they both arise by the method we have described which uses C-K ordinal formulae. The initial systems are entirely different. However, in the relation between C and C' there is an interesting analogy. In Church's method the step from C to C' is performed by means of subsidiary axioms of which the most important (Church [2], p. 88, 1_m) is almost a direct translation into his symbolism of the rule that we may take any formula of form (8.4) as an axiom. There are other extra axioms, however, in Church's system, and it is therefore not unlikely that it is in some sense more complete than Λ_P.

There are other types of ordinal logic, apparently quite unrelated to the type we have so far considered. I have in mind two types of ordinal logic, both of which can be best described directly in terms of ordinal formulae without any reference to C-K ordinal formulae. I shall describe here a specimen of one type, suggested by Hilbert (Hilbert [1], 183ff), and leave the other type over to § 12.

Suppose we have selected a particular ordinal formula $\underline{\Omega}$. We shall construct a modification $P_{\underline{\Omega}}$ of the system P of Gödel (see footnote [16]). We shall say that a natural number n is a <u>type</u> if it is either even or $2p-1$ where $\underline{\Omega}(\underline{P}, \underline{P})$ conv 3. The definition of a variable in P is to be modified by the condition that the only admissible subscripts are to be the types in our sense. Elementary expressions are then defined as in P: in particular the definition of an elementary expression of type 0 is unchanged. An elementary formula is defined to be a sequence of symbols of the form $\mathfrak{A}_m \mathfrak{A}_n$ where $\mathfrak{A}_m$, $\mathfrak{A}_n$ are elementary expressions of types m, n satisfying one of the conditions (a), (b), (c).

(a) m and n are both even and m exceeds n,

(b) m is odd and n is even,

(c) $m = 2p-1$, $n = 2q-1$ and $\underline{\Omega}(p, q)$ conv 1.

With these modifications the formal development of $P_{\underline{\Omega}}$ is the same as that of P. We wish however to have a method of associating number theoretic theorems with certain of the formulae of $P_{\underline{\Omega}}$. We cannot take over directly the association we used in P. Suppose G is a formula in P interpretable as a number theoretic theorem in the way we described when constructing Λ_P (p. 50). Then if every type suffix in G is doubled we shall obtain a formula in $P_{\underline{\Omega}}$ which is to be interpreted as the same number theoretic theorem. By the method of § 6 we can now obtain from $P_{\underline{\Omega}}$ a formula $L_{\underline{\Omega}}$ which is a logic formula of $P_{\underline{\Omega}}$ is valid; in fact given $\underline{\Omega}$ there is a method of obtaining $L_{\underline{\Omega}}$, so that there is a formula Λ_H such that $\Lambda_H(\underline{\Omega})$ conv $L_{\underline{\Omega}}$ for each ordinal formula $\underline{\Omega}$.

Having now familiarized ourselves with ordinal logics by means of these examples we may begin to consider general questions concerning them.

9. Completeness questions.

The purpose of introducing ordinal logics was to avoid as far as possible the effects of Gödel's theorem. It is a consequence of this theorem, suitably modified, that it is impossible to obtain a complete logic formula, or (roughly speaking now) a complete system of logic. We were able, however, from a given system to obtain a more complete one by the adjunction as axioms of formulae, seen intuitively to be correct, but which the Gödel theorem shows are unprovable[21] in the original system; from this we obtained a yet more complete system by a repetition of the process, and so on. We found that the repetition of the process gave us a new system for each C-K ordinal formula. We should like to know whether this process suffices, or whether the system should be extended in other ways as well. If it were possible to tell of a W.F.F. in normal form whether it was an ordinal formula we should know for certain that it was necessary to extend in other ways. In fact for any ordinal formula $\underline{\Lambda}$ it would then be possible to find a single logic formula $\underline{L}$ such that if $\underline{\Lambda}(\underline{\Omega}, \underline{A})$ conv 2 for some ordinal formula $\underline{\Omega}$ then $\underline{L}(\underline{A})$ conv 2. Since $\underline{L}$ must be incomplete there must be formulae $\underline{A}$ for which $\underline{\Lambda}(\underline{\Omega}, \underline{A})$ is not convertible to 2 for any ordinal formula $\underline{\Omega}$. However, in view of the fact, proved in § 7, that there is no method of determining of a formula in normal form whether it is an ordinal formula, the case does not arise, and there is still a possibility that some

---

[21] In the case of P we adjoin all of the axioms $(\exists x_0)\,\mathrm{Proof}_P[x_0, f^{(m)}0]$, where m is the G.R. of $\mathfrak{F}$, some of which the Gödel theorem shows to be unprovable in P.

ordinal logics may be complete in some sense. There is quite a natural way of defining completeness.

Definition of completeness of an ordinal logic. We say that an ordinal logic $\underline{\Lambda}$ is complete if for each dual formula $\underline{A}$ there is an ordinal formula $\underline{\Omega}_{\underline{A}}$ such that $\underline{\Lambda}(\underline{\Omega}_{\underline{A}}, \underline{A})$ conv 2.

As has been explained in § 2, the reference in the definition to the existence of $\underline{\Omega}_{\underline{A}}$ for each $\underline{A}$ is to be understood in the same naive way as any reference to existence in mathematics.

There is room for modification in this definition: we might require that there be a formula $\underline{X}$ such that $\underline{X}(\underline{A})$ conv $\underline{\Omega}_{\underline{A}}$, $\underline{X}(\underline{A})$ being an ordinal formula whenever $\underline{A}$ is dual. There is no need, however, to discuss the relative merits of these two definitions, because in all cases where we prove an ordinal logic to be complete we shall prove it to be complete even in the modified sense, but in cases where we prove an ordinal logic to be incomplete we use the definition as it stands.

In the terminology of § 6 $\underline{\Lambda}$ is complete if the class of logics $\underline{\Lambda}(\underline{\Omega})$ is complete when $\underline{\Omega}$ runs through all ordinal formulae.

There is another completeness property which is related to this one. Let us for the moment say that an ordinal logic $\underline{\Lambda}$ is all inclusive if to each logic formula $\underline{L}$ there corresponds an ordinal formula $\underline{\Omega}_{(\underline{L})}$ such that $\underline{\Lambda}(\underline{\Omega}_{(\underline{L})})$ is as complete as $\underline{L}$. Clearly every all inclusive ordinal logic is complete, for if $\underline{A}$ is dual then $\delta(\underline{A})$ is a logic with $\underline{A}$ in its extent. But if $\underline{\Lambda}$ is complete and

▷ $A i \to \lambda k w a. \Gamma(\lambda r. S(4, S(2, k(w, V(Nm(r)))) + S(2, Nm(r, a)))$

then $Ai(\underline{\Lambda})$ is an all inclusive ordinal logic. For if $\underline{A}$ is in the extent of $\underline{\Lambda}(\underline{\Omega}_{\underline{A}})$ for each $\underline{A}$, and we put $\underline{\Omega}_{(\underline{L})} \to \underline{\Omega}_{V(\underline{L})}$ then I say that if $\underline{B}$ is in the extent of $\underline{L}$ it must be in the extent of $Ai(\underline{\Lambda}, \underline{\Omega}_{(\underline{L})})$. In fact $Ai(\underline{\Lambda}, \underline{\Omega}_{V(\underline{L})}, \underline{B})$ conv $\Gamma(\lambda r. \delta(4, \delta(2, \underline{\Lambda}(\underline{\Omega}_{V(\underline{L})}, V(Nm(r)))) + \delta(2, Nm(r, \underline{B}))))$

For suitable n, $Nm(\underline{n})$ conv $\underline{L}$ and then

$$\underline{\Lambda}(\underline{\Omega}_{V(\underline{L})}, V(Nm(\underline{n}))) \quad \text{conv } 2$$

$$Nm(\underline{n}, \underline{B}) \quad \text{conv } 2$$

and therefore by the properties of Γ, δ

$$Ai(\underline{\Lambda}, \underline{\Omega}_{V(\underline{L})}, \underline{B}) \quad \text{conv } 2$$

Conversely $Ai(\underline{\Lambda}, \underline{\Omega}_{V(\underline{L})}, \underline{B})$ can only be convertible to 2 if both $Nm(\underline{n}, \underline{B})$ and $\underline{\Lambda}(\underline{\Omega}_{V(\underline{L})}, V(Nm(\underline{n})))$ are convertible to 2 for some positive integer n; but if $\underline{\Lambda}(\underline{\Omega}_{V(\underline{L})}, V(Nm(\underline{n})))$ conv 2 then $Nm(\underline{n})$ must be a logic and since $Nm(\underline{n}, \underline{B})$ conv 2, $\underline{B}$ must be dual.

It should be noticed that our definitions of completeness refer only to number theoretic theorems. Although it would be possible to introduce formulae analogous to ordinal logics which would prove more general theorems than number theoretic ones, and have a corresponding definition of completeness, yet if our theorems are too general we shall find that our (modified) ordinal logics are never complete. This follows from the argument of § 4. If our 'oracle' tells us, not whether any given number theoretic statement is true, but whether a given formula is an ordinal formula, or the argument still applies, and we find there are classes of problem which cannot be solved by a uniform process even with the help of this oracle. This is equivalent to saying that there is no ordinal logic of the proposed modified type which is complete with respect to these problems. This situation becomes more definite if we take formulae satisfying conditions (a) - (e), (f') (as described at the end of §12) instead of ordinal formulae; it is then not possible for the ordinal logic to be complete with respect to any class of problems more extensive than the number theoretic problems.

We might hope to obtain some intellectually satisfying system of logical inference (for the proof of number theoretic theorems) with some ordinal logic. Gödel's theorem shows that such a system cannot be wholly mechanical, but with a complete ordinal logic we should be able to confine the non-mechanical steps entirely to verifications that particular formulae are ordinal formulae.

We might also expect to obtain an interesting classification of number theoretic theorems according to 'depth'. A theorem which required an ordinal α to prove it would be deeper than one which could be proved by the use of an ordinal β less than α. However, this presupposes more than is justified. We define

Definition of invariance of ordinal logics. An ordinal logic $\underline{\Lambda}$ is said to be invariant up to an ordinal α if, whenever $\underline{\Omega}$, $\underline{\Omega}'$ are ordinal formulae representing the same ordinal less than α, the extent of $\underline{\Lambda}(\underline{\Omega})$ is identical with the extent of $\underline{\Lambda}(\underline{\Omega}')$. An ordinal logic is invariant if it is invariant up to each ordinal represented by an ordinal formula.

Clearly the classification into depths presupposes that the ordinal logic used is invariant.

Among the questions we should now like to ask are

(a) are there any complete ordinal logics?

(b) are there any complete invariant ordinal logics?

To these we might have added 'are all ordinal logics complete?'; but this is trivial; in fact there are ordinal logics which do not suffice to prove any number theoretic theorems whatever.

We shall now show that (a) must be answered affirmatively. In fact we can write down a complete ordinal logic at once. Put

$$Od \rightarrow \lambda a.\{\lambda fmn.\mathfrak{D}t(f(m), f(n))\}(\lambda s.\mathfrak{P}(\lambda r.r(\underline{I}, a(s)), 1, s))$$

and

$$Comp \rightarrow \lambda \omega a.\,\delta(\omega, Od(a))$$

I shall show that $Comp$ is a complete ordinal logic.

In ~~fact~~ For if $Comp(\underline{\Omega}, \underline{A})$ conv 2, then

$\underline{\Omega}$ conv $Od(\underline{A})$

conv $\lambda mn.\mathfrak{D}t(\mathfrak{P}(\lambda r.r(\underline{I}, \underline{A}(m)), 1, m)), \mathfrak{P}(\lambda r.r(\underline{I}, \underline{A}(n)), 1, n))$

$\underline{\Omega}(\underline{m}, \underline{n})$ has a normal form if $\underline{\Omega}$ is an ordinal formula, so that then $\mathfrak{P}(\lambda r.r(\underline{I}, \underline{A}(\underline{m})), 1)$ has a normal form; this means that $\underline{r}(\underline{I}, \underline{A}(\underline{m}))$ conv 2 some r, i.e. $\underline{A}(\underline{m})$ conv 2. Thus if $Comp(\underline{\Omega}, \underline{A})$ conv 2 and $\underline{\Omega}$ is an ordinal formula then $\underline{A}$ is dual. $Comp$ is therefore an ordinal logic. Now suppose conversely that $\underline{A}$ is dual. I shall show that $Od(\underline{A})$ is an ordinal formula representing the ordinal ω. In ~~fact~~ For

$$\mathfrak{P}(\lambda r.r(\underline{I}, \underline{A}(\underline{m})), 1, \underline{m}) \text{ conv } \mathfrak{P}(\lambda r.r(\underline{I}, 2), 1, \underline{m})$$

$$\text{conv } I(\underline{m}) \text{ conv } \underline{m}$$

$$Od(\underline{A}, \underline{m}, \underline{n}) \text{ conv } Dt(\underline{m}, \underline{n})$$

i.e. $Od(\underline{A})$ is an ordinal formula representing the same ordinal as Dt. But

$$Comp(Od(\underline{A}), \underline{A}) \text{ conv } \delta(Od(\underline{A}), Od(\underline{A})) \qquad \text{conv } 2$$

This proves the completeness of $Comp$.

Of course $Comp$ is not the kind of complete ordinal logic that we should really want to use. The use of $Comp$ does not make it any easier to see that $\underline{A}$ is dual. ~~In~~-fact if we really want to use an ordinal logic a proof of completeness for that particular ordinal logic will be of little value; the ordinals given by the completeness proof will not be ones which can ~~really~~ easily be seen intuitively to be ordinals. The only value ~~of~~ a completeness proof of this kind would have would be to show that if any objection is to be raised against an ordinal logic it must be on account of something more subtle then incompleteness.

The theorem of completeness is also unexpected in that the ordinal formulae used are all formulae representing ω. This is contrary to our intentions in constructing Λ_P for instance; implicitly we had in mind large ordinals expressed in a simple manner Here we have small ordinals expressed in a very complex and artificial way.

Before trying to solve the problem (b), let us see how far Λ_P and Λ_T are invariant. We should certainly not expect Λ_P to be invariant, as the extent of $\Lambda_P(\underline{\Omega})$ will depend on whether $\underline{\Omega}$

is convertible to a formula of form $H(\underline{A})$: but suppose we call an ordinal logic $\underline{\Lambda}$ C-K invariant up to α if the extent of $\underline{\Lambda}(H(\underline{A}))$ is the same as the extent of $\underline{\Lambda}(H(\underline{B}))$ whenever $\underline{A}$ and $\underline{B}$ are C-K ordinal formulae representing the same ordinal less than α. How far is Λ_P C-K invariant? It is not difficult to see that it is C-K invariant up to any finite ordinal, that is to say up to ω. It is also C-K invariant up to $\omega+1$, and follows from the fact that the extent of $\Lambda_P(H(\lambda u f x . u(\underline{B})))$ is the set theoretic sum of the extents of

$$\Lambda_P(H(\lambda u f x . \underline{R}(1))), \Lambda_P(H(\lambda u f x . \underline{R}(2))), \ldots$$

However, there is no obvious reason to believe that it is C-K invariant up to $\omega+2$, and in fact it is demonstrable that this is not the case (see the end of this section). Let us try to see what happens if we try to prove that the extent of $\Lambda_P(H(Suc(\lambda u f x . u(\underline{B}_1))))$ is the same as the extent of $\Lambda_P(H(Suc(\lambda u f x . u(\underline{R}_2))))$ where $\lambda u f x . u(\underline{R}_1)$ and $\lambda u f x . u(\underline{B}_2)$ are two C-K ordinal formulae representing ω. We should have to prove that a formula interpretable as a theorem of number theory is provable in $C[Suc(\lambda u f x . u(\underline{B}_1))]$ if and only if it is provable in $C[Suc(\lambda u f x . u(\underline{R}_2))]$. Now $C[Suc(\lambda u f x . u(\underline{B}_1))]$ is obtained from $C[\lambda u f x . u(\underline{R}_1)]$ by adjoining all axioms of form

$$(\exists x_0) Proof_{C[\lambda u f x . u(R_1)]}[x_0, f^{(m)}0] \supset \mathcal{F} \qquad (9.1)$$

where m is the G.R. of $\mathcal{F}$, and $C[Suc(\lambda u f x . u(\underline{B}_2))]$ is obtained from $C[\lambda u f x . u(\underline{B}_2)]$ by adjoining all axioms of form

$$(\exists x_0) Proof_{C[\lambda u f x . u(\underline{B}_2)]}[x_0, f^{(m)}0] \supset \mathcal{F} \qquad (9.2)$$

The axioms which must be adjoined to P to obtain $C[\lambda u f x . u(\underline{R}_1)]$ are essentially the same as those which must be adjoined to obtain $C[\lambda u f x . u(\underline{R}_2)]$: however the *rules of procedure which have to be applied before these axioms can be written down will in general be quite different in the two cases.* Consequently (9.1) and (9.2) will be quite different axioms, and there is no reason to expect their consequences to be the same. A proper understanding of this will make our treatment of question (b) much more intelligible. See also footnote .

Now let us turn to Λ_T. This ordinal logic is invariant. Suppose $\underline{\Omega}$, $\underline{\Omega}'$ represent the same ordinal, and suppose we have a proof of a number theoretic theorem G in $P_{\underline{\Omega}}$. The formula expressing the number theoretic theorem does not involve any odd types. Now there is a one-one correspondence between the odd types such that if $2m-1$ corresponds to $2m'-1$ and $2n-1$ to $2n'-1$ then $\underline{\Omega}(\underline{m}, \underline{n})$ conv 2 implies $\underline{\Omega}'(\underline{m}', \underline{n}')$ conv 2. Let us modify the odd type-subscripts occurring in the proof of G, replacing each by its mate in the one-one correspondence. There results a proof in $P_{\underline{\Omega}'}$ with the same end formula G. That is to say that if G is provable in $P_{\underline{\Omega}}$ it is provable in $P_{\underline{\Omega}'}$; Λ_T is invariant.

The question (b) must be answered negatively. Much more can be proved, but we shall first prove an even weaker result which can be established very quickly, in order to illustrate the method.

I shall prove that an ordinal logic $\underline{\Lambda}$ cannot be invariant and have the property that the extent of $\underline{\Lambda}(\underline{\Omega})$ is a strictly increasing function of the ordinal represented by $\underline{\Omega}$. Suppose $\underline{\Lambda}$ has these properties; we shall obtain a contradiction. Let $\underline{A}$ be a W.F.F. in normal form and without free variables, and consider the process of carrying out conversions on $\underline{A}(1)$ until we have shown it convertible to 2, then converting $\underline{A}(2)$ to 2, then $\underline{A}(3)$ and so on: suppose that after r steps we are still performing the conversion on $\underline{A}(\underline{m}_r)$. There is a formula Jh such that $Jh(\underline{A}, \underline{r})$ conv $\underline{m}_r$ for each positive integer r . Now let Z be a formula such that for each positive integer n , $Z(\underline{n})$ is an ordinal formula representing ω^n , and suppose $\underline{B}$ is a member of the extent of $\underline{\Lambda}(Suc(Lim(Z)))$ but not of the extent of $\underline{\Lambda}(Lim(Z))$. Put

$$\underline{K}^* \to \lambda a.\, \underline{\Lambda}(Suc(Lim(\lambda r.\, Z(Jh(a, r)))), \underline{B})$$

then $\underline{K}^*$ is a complete logic. For if $\underline{A}$ is dual, then $Suc(Lim(\lambda r.\, Z(Jh(\underline{A}, r))))$ represnnts the ordinal $\omega^\omega + 1$, and therefore $\underline{K}^*(\underline{A})$ conv 2; but if $\underline{A}(\underline{u})$ is not convertible to 2, then $Suc(Lim(\lambda r.\, Z(Jh(\underline{A}, r))))$ represents an ordinal not exceeding $\omega^u + 1$, and $\underline{K}^*(\underline{A})$ is therefore not convertible to 2. Since there are no complete logic formulae this proves our assertion.

We may now prove more powerful results.

Incompleteness theorems. (A) If an ordinal logic $\underline{\Lambda}$ is invariant up to an ordinal α , then for any ordinal formula $\underline{\Omega}$ representing an ordinal β , $\beta < \alpha$, the extent of $\underline{\Lambda}(\underline{\Omega})$ is contained in the (set-theoretic) sum of the extents of the logics $\underline{\Lambda}(P)$ where P is finite.

(B) If an ordinal logic $\underline{\Lambda}$ is C-K invariant up to an ordinal α, then for any C-K ordinal formula $\underline{A}$ representing an ordinal β, $\beta<\alpha$, the extent of $\underline{\Lambda}(H(\underline{A}))$ is contained in the (set-theoretic) sum of the extents of the logics $\Lambda(H(F))$ where $\underline{F}$ is a C-K ordinal formula representing an ordinal less than ω^2.

Proof of (A). It suffices to prove that if $\underline{\Omega}$ represents an ordinal γ, $\omega \leq \gamma < \alpha$, then the extent of $\underline{\Lambda}(\underline{\Omega})$ is contained in the set theoretic sum of the extents of the logics $\underline{\Lambda}(\underline{\Omega}')$ where $\underline{\Omega}'$ represents an ordinal less than γ. The ordinal γ must be of the form $\gamma_0+\rho$ where ρ is finite and represented by $\underline{P}$ say, and γ_0 is not the successor of any ordinal and is not less than ω. There are two cases to consider; $\gamma_0 = \omega$ and $\gamma_0 \geqslant 2\omega$. In each of them we shall obtain a contradiction from the assumption that there is a W.F.F. $\underline{B}$ such that $\underline{\Lambda}(\underline{\Omega}, \underline{B})$ conv 2 whenever $\underline{\Omega}$ represents γ, but is not convertible to 2 if $\underline{\Omega}$ represents a smaller ordinal. Let us take first the case $\gamma_0 \geqslant 2\omega$. Suppose $\gamma_0 = \omega+\gamma_1$, and that $\underline{\Omega}_1$ is an ordinal formula representing γ_1. Let $\underline{A}$ be any W.F.F. with a normal form and no free variables, and let Z be the class of those positive integers which are exceeded by all integers n for which $\underline{A}(\underline{n})$ is not convertible to 2. Let E be the class of integers $2p$ such that $\underline{\Omega}(\underline{P}, \underline{n})$ conv 2 for some n belonging to Z. The class E, together with the class φ of all odd integers is constructively enumerable. It is evident that the class can be enumerated with repetitions, and since it is infinite the required enumeration can be obtained by striking out the repetitions. There is, therefore, a formula $\underline{Ew}$ such that $\underline{Ew}(\underline{\Omega}, \underline{A}, \underline{r})$ runs through the formulae of the class $E + \varphi$ without repetitions as r runs through the positive integers. We define

$$Rt \rightarrow \lambda \omega a m n . Sum(Dt, \omega, En(\omega, a, m), En(\omega, a, n))$$

Then $Rt(\underline{\Omega}_1, \underline{A})$ is an ordinal formula which represents γ_0 if $\underline{A}$ is dual, but a smaller ordinal otherwise. In fact

$$Rt(\underline{\Omega}_1, \underline{A}, \underline{m}, \underline{n}) \text{ conv } \{Sum(Dt, \underline{\Omega}_1)\}(En(\underline{\Omega}_1, \underline{A}, \underline{m}), En(\underline{\Omega}_1, \underline{A}, \underline{n}))$$

Now if $\underline{A}$ is dual $E + \varphi$ includes all integers m for which $\{Sum(Dt, \underline{\Omega}_1)\}(\underline{m}, \underline{m})$ conv 2. Putting "$En(\underline{\Omega}_1, \underline{A}, \underline{P})$ conv q" for $M(p, q)$ we see that condition (7.4) is satisfied, so that $Rt(\underline{\Omega}_1, \underline{A})$ is an ordinal formula representing γ_0. But if $\underline{A}$ is not dual the set $E + \varphi$ consists of all integers m for which $\{Sum(Dt, \underline{\Omega}_1)\}(\underline{m}, \underline{r})$ conv 2, where r depends only on $\underline{A}$. In this case $Rt(\underline{\Omega}_1, \underline{A})$ is an ordinal formula representing the same ordinal as $Inf(Sum(Dt, \underline{\Omega}_1), \underline{r})$, and this is smaller than γ_0. Now consider $\underline{K}$:

$$\underline{K} \rightarrow \lambda a . \Lambda(Sum(Rt(\underline{\Omega}_1, \underline{A}), \underline{P}), \underline{B})$$

If $\underline{A}$ is dual, $\underline{K}(\underline{A})$ is convertible to 2, since $Sum(Rt(\underline{\Omega}_1, \underline{A}), \underline{P})$ represents γ. But if $\underline{A}$ is not dual it is not convertible to 2, for $Sum(Rt(\underline{\Omega}_1, \underline{A}), \underline{P})$ then represents an ordinal smaller than γ. In K we therefore have a complete logic formula, which is impossible.

Now we take the case $\gamma_0 = \omega$. We introduce a W.F.F. Mg such that if n is the D.N. of a computing machine $\mathcal{M}$, and if by the mth complete configuration of $\mathcal{M}$ the figure 0 has been printed then $Mg(\underline{n}, \underline{m})$ is convertible to $\lambda p q . Al(4(P, 2p + 2q), 3, 4)$ (which is an ordinal formula representing the ordinal 1), but if 0 has not been printed it is convertible to $\lambda p q . p(q, \underline{I}, 4)$

(which represents 0). Now consider $\underline{M}$.

$$\underline{M} \rightarrow \lambda n.\, \underline{\Lambda}\left(Sum\left(lim\left(Mg(n)\right), \underline{P}\right), \underline{B}\right)$$

If the machine never prints 0 then $lim(\lambda r.\, Mg(n,r))$ represents ω and $Sum(lim(Mg(\underline{n})),\underline{P})$ represents γ. This means that $Mg(\underline{n})$ is convertible to 2. If, however, $\mathcal{M}$ never prints 0, $Sum(lim(Mg(\underline{n})),\underline{P})$ represents a finite ordinal and $\underline{M}(\underline{n})$ is not convertible to 2. In $\underline{M}$ we therefore have a means of determining of a machine whether it ever prints 0, which is impossible[22]. (Turing [1], § 8). This completes the proof of (A).

[22] This part of the argument can equally well be based on the impossibility of determining of two W.F.F. whether they are interconvertible. (Church [3], 363.)

<u>Proof of (B)</u>. It suffices to prove that if $\underline{C}$ represents an ordinal γ, $\omega^\omega \leq \gamma < \alpha$ then the extent of $\underline{\Lambda}(H(\underline{C}))$ is included in the set-theoretic sum of the extents of $\underline{\Lambda}(H(\underline{G}))$ where $\underline{G}$ represents an ordinal less than γ. We obtain a contradiction from the assumption that there is a formula $\underline{B}$ which is in the extent of $\underline{\Lambda}(H(\underline{G}))$ if $\underline{G}$ represents γ, but not if it represents any smaller ordinal. The ordinal γ is of the form $\delta + \omega^\omega + \xi$ where $\xi < \omega^\omega$. Let $\underline{D}$ be a C-K ordinal formula representing δ and $\underline{\varphi}$ one representing ξ.

We now define a formula Hg. Suppose $\underline{A}$ is a W.F.F. in normal form and without free variables; consider the process of carrying out conversions on $\underline{A}(\underline{1})$ until it is brought into the form 2, then converting $\underline{A}(2)$ to 2, then $\underline{A}(3)$, and so on. Suppose that at the r th step of this process we are doing the n_rth step in the conversion of $\underline{A}(\underline{m}_v)$. Thus for instance if $\underline{A}(3)$ be not convertible to 2, m_r can never exceed 3. Then $Hg(\underline{A}, \underline{r})$ is to be convertible to $\lambda f . f(\underline{m}_r, \underline{n}_r)$ for each positive integer r. Put

$$Sq \rightarrow \lambda d m n .\, n(Suc, m(\lambda a u f x .\, u(\lambda y .\, y(Suc, a, u, f, x)), d(u, f, x)))$$
$$\underline{M} \rightarrow \lambda a u f x .\, \Phi(u, f, u(\lambda y .\, Hg(a, y, Sq(\underline{D}))))$$
$$\underline{K}_1 \rightarrow \lambda a .\, \underline{\Lambda}(\underline{M}(a), \underline{B})$$

then I say that $\underline{K}_1$ is a complete logic formula. $Sq(\underline{D}, \underline{m}, \underline{n})$ is a C-K ordinal formula representing $\delta + m\omega + n$, and therefore $Hg(\underline{A}, \underline{r}, Sq(\underline{D}))$ represents an ordinal ζ_r which increases steadily with increasing r, and tends to the limit $\delta + \omega^2$ if $\underline{A}$ is dual. Further $Hg(\underline{A}, \underline{r}, Sq(\underline{D})) < Hg(\underline{A}, S(r), Sq(\underline{D}))$ for each positive integer r. $\lambda u f x .\, u(\lambda y .\, Hg(\underline{A}, y, Sq(\underline{D})))$ is therefore a C-K ordinal formula and represents the limit of the sequence $\zeta_1, \zeta_2, \zeta_3 \ldots$ This is $\delta + \omega^2$ if $\underline{A}$ is dual, but a smaller ordinal otherwise. Likewise $\underline{M}(\underline{A})$ represents γ if $\underline{A}$ is dual, but a smaller ordinal otherwise. The formula $\underline{B}$ therefore belongs to the extent of $\underline{\Lambda}(H(\underline{M}(\underline{A})))$ if and only if $\underline{A}$ is dual, and this implies that $\underline{K}_1$ is a complete logic formula as was asserted. But this is impossible and we have the required contradiction.

As a corollary to (A) we see that Λ_H is incomplete and in fact that the extent of $\Lambda_H(\underline{D}t)$ contains the extent of $\Lambda_H(\underline{\Omega})$ for any ordinal formula $\underline{\Omega}$. This result, suggested to me first by the solution of question (b), may also be obtained more directly. In fact if a number theoretic theorem can be proved in any particular $P_{\underline{\Omega}}$ it can be proved in $P_{\lambda mn.\, m(n, I, n)}$. The formulae describing number theoretic theorems in P do no involve more than a finite number of types, type 5 being the highest necessary. The formulae describing the number theoretic theorems in any $P_{\underline{\Omega}}$ will be obtained by doubling the type subscripts. Now suppose we have a proof of a number theoretic theorem G in $P_{\underline{\Omega}}$ and that the types occurring in the proof are among 0, 2, 4, 6, 8, 10, $t_1, t_2, t_3, \ldots t_R$. We may suppose they have been arranged with all the even types preceding all the odd types, the even types in order of magnitude and the type $2m-1$ preceding $2n-1$ if $\underline{\Omega}(\underline{m}, \underline{n})$ conv 2. Now let each t_r be replaced by $10+2r$ throughout the proof of G. We obtain a proof of G in $P_{\lambda mn.\, m(n, I, 4)}$.

As with problem (a) the solution of problem (b) does not require the use of high ordinals (e.g. if we make the assumption that the extent of $\underline{\Lambda}(\underline{\Omega})$ is a steadily increasing function of the ordinal represented by $\underline{\Omega}$ we do not have to consider ordinals higher than $2\omega+2$). However, if we restrict what we are to call ordinal formulae in some way we shall have corresponding modified problems (a) and (b); the solutions will presumably be essentially the same but will involve higher ordinals. Suppose for example that $Prod$ is a W.F.F. with the property that $Prod(\underline{\Omega}_1, \underline{\Omega}_2)$ is an ordinal formula representing $\alpha_1 \alpha_2$ when $\underline{\Omega}_1, \underline{\Omega}_2$ are ordinal formulae representing α_1, α_2 respectively and suppose we call a W.F.F. a

1-ordinal formula when it is convertible to the form $\text{Sum}(\text{Prod}(\underline{Q}, \text{Dt}), \underline{P})$ where $\underline{Q}$, $\underline{P}$ are ordinal formulae of which $\underline{P}$ represents a finite ordinal. We may define 1-ordinal logics, 1-completeness and 1-invariance in an obvious way, and obtain a solution of problem (b) which differs from the solution in the ordinary case in that the ordinals less than ω take the place of the finite ordinals. More generally the cases I have in mind will be covered by the following theorem.

Suppose we have a class V of formulae representing ordinals in some manner we do not propose to specify definitely, and a subset[23] U of the class V such that

(i) There is a formula $\underline{\underline{\Phi}}$ such that if $\underline{T}$ enumerates a sequence of members of U representing an increasing sequence of ordinals, then $\underline{\underline{\Phi}}(\underline{T})$ is a member of U representing the limit of the sequence.

(ii) There is a formula $\underline{E}$ such that $\underline{E}(\underline{m}, \underline{n})$ is a member of U for each pair of positive integers m, n and if it represents $\varepsilon_{m,n}$ then $\varepsilon_{m,n} < \varepsilon_{m',n'}$ if either $m < m'$ or $m = m'$, $n < n'$

(iii) There is a formula $\underline{G}$ such that if $\underline{A}$ is a member of U then $\underline{G}(\underline{A})$ is a member of U representing a larger ordinal than does $\underline{A}$, and such that $\underline{G}(\underline{E}(\underline{m}, \underline{n}))$ always represents an ordinal not larger than $\varepsilon_{m,n+1}$.

We define a V-ordinal logic to be a W.F.F. $\underline{\Lambda}$ such that $\underline{\Lambda}(\underline{A})$ is a logic whenever $\underline{A}$ belongs to V. $\underline{\Lambda}$ is V-invariant if the extent

---

[23] The subset U wholly supersedes V in what follows. The introduction of V serves to emphasise the fact that the set of ordinals represented by members of U may have gaps.

of $\underline{\Lambda}(\underline{A})$ depends only on the ordinal represented by $\underline{A}$. Then it is not possible for a V-ordinal logic $\underline{\Lambda}$ to be V-invariant and have the property that if $\underline{C}_1$ represents a greater ordinal than $\underline{C}_2$, ($\underline{C}_1$ and $\underline{C}_2$ both being members of U) then the extent of $\underline{\Lambda}(\underline{C}_1)$ is greater than the extent of $\underline{\Lambda}(\underline{C}_2)$.

We suppose the contrary. Let $\underline{B}$ be a formula belonging to the extent of $\underline{\Lambda}(\underline{G}(\underline{\Phi}(\lambda r.\,\underline{E}(r, 1))))$ but not to the extent of $\underline{\Lambda}(\underline{\Phi}(\lambda r.\,\underline{E}(r, 1)))$. Suppose that our assertion is false and that

$$\underline{K}' \rightarrow \lambda a.\,\underline{\Lambda}(\underline{\Theta}(\lambda r.\,\mathrm{Hg}(a, r, \underline{E})), \underline{B})$$

Then $\underline{K}'$ is a complete logic. For

$$\mathrm{Hg}(\underline{A}, \underline{r}, \underline{E}) \quad \text{conv} \quad \underline{E}(\underline{m}_r, \underline{n}_r)$$

$\underline{E}(\underline{m}_r, \underline{n}_r)$ is a sequence of V-ordinal formulae representing an increasing sequence of ordinals. Their limit is represented by $\underline{\Theta}(\lambda r.\,\mathrm{Hg}(\underline{A}, r, \underline{E}))$; let us see what this limit is. First suppose $\underline{A}$ is dual: then m_r tends to infinity as r tends to infinity, and $\underline{\Theta}(\lambda r.\,\mathrm{Hg}(\underline{A}, r, \underline{E}))$ therefore represents the same ordinal as $\underline{\Theta}(\lambda r.\,\underline{E}(r, 1))$. In this case we must have $\underline{K}'(\underline{A})$ conv 2. Now suppose $\underline{A}$ is not dual: m_r is eventually equal to some constant number, a say, and $\underline{\Theta}(\lambda r.\,\mathrm{Hg}(\underline{A}, r, \underline{E}))$ represents the same ordinal as $\underline{\Theta}(\lambda r.\,\underline{E}(\underline{a}, r))$ which is smaller than that represented by $\underline{\Theta}(\lambda r.\,\underline{E}(r, 1))$. $\underline{B}$ cannot therefore belong to the extent of $\underline{\Theta}(\lambda r.\,\mathrm{Hg}(\underline{A}, r, \underline{E}))$, and $\underline{K}(\underline{A})$ is not convertible to 2. We have proved that $\underline{K}'$ is a complete logic which is impossible.

This theorem can no doubt be improved in many ways. However, it is sufficiently general to show that, with almost any reasonable notation for ordinals, completeness is incompatible with invariance.

We can still give a certain meaning to the classification into depths with highly restricted kinds of ordinals. Suppose we take a particular ordinal logic $\underline{\Lambda}$ and a particular ordinal formula $\underline{\Psi}$ representing the ordinal α say (preferably a large one), and we restrict ourselves to ordinal formulae of the form $Inf(\underline{\Psi}, \underline{a})$. We shall then have a classification into depths, but the extents of all the logics we so obtain will be contained in the extent of a single logic.

We now attempt a problem of a rather different character, that of the completeness of Λ_P. It is to be expected that this ordinal logic is complete. I cannot at present give a proof of this, but I can give a proof that it is complete as regards a simpler type of theorem than the number theoretic theorems viz. those of form '$\theta(x)$ vanishes identically' where $\theta(x)$ is primitive recursive. The proof will have to be much abbreviated as we do not wish to go into the formal details of the system P. Also there is a certain lack of definiteness in the problem as at present stated, owing to the fact that the formulae G, E, M_P were not completely defined. Our attitude here is that it is open to the sceptical reader to give detailed definitions for these formulae and then verify that the remaining details of the proof can be filled in using his definition. It is not asserted that these details can be filled in whatever be the definitions of G, E, M_P consistent with the properties already required of them, only that it is so with the more natural definitions.

I shall prove the completeness theorem in the following form. If $\mathfrak{G}[x_0]$ is a recursion formula and $\mathfrak{G}[0]$, $\mathfrak{G}[f0]$, . . . are all provable in P, then there is a C-K ordinal formula $\underline{A}$ such that $(x_0)\mathfrak{G}[x_0]$ is provable in the system $P^{\underline{A}}$ of logic obtained from P by adjoining as axioms all formulae whose G.R's are of the form

$$\underline{A}(\lambda mn.m(\varpi(2,n),\varpi(3,n)),K,M_P,\underline{r})$$

(provided they represent propositions.)

First let us define the formula $\underline{A}$. Suppose $\underline{D}$ is a W.F.F. with the property that $\underline{D}(\underline{n})$ conv 2 if $\mathfrak{G}[f^{(n-1)}0]$ is provable in P, but $\underline{D}(\underline{n})$ conv 1 if $\sim\mathfrak{G}[f^{(n-1)}0]$ is provable in P (P is being assumed consistent). Let Θ be defined by

$$\Theta \rightarrow \{\lambda vu.v(v(v,u))\}(\lambda vu.v(v(v,u)))$$

and let V be a formula with the properties

$$V(2) \text{ conv } \lambda u.u(Suc,U)$$

$$V(1) \text{ conv } \lambda u.u(\underline{I},\Theta(Suc))$$

The existence of such a formula is established in Kleene 1, corollary on p 220. Now put

$$\underline{A}^* \rightarrow \lambda ufx.u(\lambda y.V(\underline{D}(y),y,u,f,x))$$

$$A \rightarrow Suc(\underline{A}^*)$$

I assert that $\underline{A}^*$, $\underline{A}$ are C-K ordinal formulae whenever it is true that $\mathfrak{B}[0]$, $\mathfrak{B}[f0]$, . . . are all provable in P. For in this case $\underline{A}^*$ is $\lambda ufx . u(\underline{R})$ where

$$\underline{R} \rightarrow \lambda y. V(\underline{D}(y), y, u, f, x)$$

and then

$$\lambda ufx . \underline{R}(\underline{n}) \text{ conv } \lambda ufx . V(\underline{D}(\underline{n}), \underline{n}, u, f, x)$$

$$\text{conv } \lambda ufx . V(2, \underline{n}, u, f, x)$$

$$\text{conv } \lambda ufx . \{\lambda u . u(Suc, U)\}(\underline{n}, u, f, x)$$

$$\text{conv } \lambda ufx . \underline{n}(Suc, U, u, f, x)$$ which is a C-K ordinal formula, and

$$\lambda ufx . S(\underline{n}, Suc, U, u, f, x) \text{ conv } Suc(\lambda ufx . \underline{n}(Suc, U, u, f, x))$$

These relations hold for an arbitrary positive integer n and therefore $\underline{A}^*$ is a C-K ordinal formula (condition (9) p. 32): it follows immediately that $\underline{A}$ is also a C-K ordinal formula. It remains to prove that $(x_0)\mathfrak{B}[x_0]$ is provable in $P_{\underline{A}}$. To do this it is necessary to examine the structure of $\underline{A}^*$ in the case that $(x_0)\mathfrak{B}[x_0]$ is false. Let us suppose that $\sim\mathfrak{B}[f^{(a-1)}0]$ is true so that $\underline{D}(\underline{a})$ conv 1, and let us consider $\underline{B}$ where

$$\underline{B} \rightarrow \lambda ufx . V(\underline{D}(\underline{a}), \underline{a}, u, f, x)$$

If $\underline{A}^*$ were a C-K ordinal formula then $\underline{B}$ would be a member of its fundamental sequence; but

$$\underline{B} \quad \text{conv} \quad \lambda u f x \,.\, V(\underline{1}, \underline{a}, u, f, x)$$

$$\text{conv} \quad \lambda u f x \,.\, \{\lambda n \,.\, n(\underline{I}, \Theta(Suc))\}(\underline{a}, u, f, x$$

$$\text{conv} \quad \lambda u f x \,.\, \Theta(Suc, u, f, x)$$

$$\text{conv} \quad \lambda u f x \,.\, \{\lambda u \,.\, u(\Theta(u))\}(Suc, u, f, x)$$

$$\text{conv} \quad \lambda u f x \,.\, Suc(\Theta(Suc), u, f, x)$$

$$\text{conv} \quad Suc(\lambda u f x \,.\, \Theta(Suc, u, f, x))$$

$$\text{conv} \quad Suc(\underline{B}) \qquad (9.3)$$

This of course implies that $\underline{B} < \underline{B}$ and therefore that $\underline{B}$ is no C-K ordinal formula. This, although fundamental to the possibility of proving our completeness theorem does not form an actual step in the argument. Roughly speaking our argument will amount to this. The relation (9.3) implies that the system $P^{\underline{B}}$ is inconsistent and therefore that $P^{\underline{A}^*}$ is inconsistent, and indeed we can prove in P (and a fortiori in $P^{\underline{A}}$) that $\sim(x_0) \,\&\, [x_0$ implies the inconsistency of $P^{\underline{A}^*}$. On the other hand in $P^{\underline{A}}$ we can prove the consistency of $P^{\underline{A}^*}$. The inconsistency of $P^{\underline{B}}$ is proved by the Godel argument. Let us return to the details.

The axioms in $P^{\underline{B}}$ are those whose G.R's are of the form

$$\underline{B}\,(\lambda m n \,.\, m(\varpi(2, n), \varpi(3, n)), K, M_p, \underline{r})$$

Replacing $\underline{B}$ by $Suc(\underline{B})$ this becomes

$$Suc(\underline{B}, \lambda mn.\, m(\mathfrak{d}(2,n), \mathfrak{d}(3,n)), K, M_p, \underline{r})$$

$$\text{conv } K(\underline{B}(\lambda mn.\, m(\mathfrak{d}(2,n), \mathfrak{d}(3,n)), K, M_p, \underline{r})$$

$$\text{conv } \underline{B}(\lambda mn.\, m(\mathfrak{d}(2,n), \mathfrak{d}(3,n)), K, M_p, \underline{r})$$

$$\text{if } \underline{r} \text{ conv } 2\underline{p}+1$$

$$\text{conv } E(\underline{B}(\lambda mn.\, m(\mathfrak{d}(2,n), \mathfrak{d}(3,n)), K, M_p), \underline{p})$$

$$\text{if } \underline{r} \text{ conv } 2\underline{p}$$

When we remember the essential property of the formula E we see that the axioms of $P_{\underline{B}}$ include all formulae of the form

$$(\exists x_0)\, Proof_{P_{\underline{B}}}[x_0, f^{(q)}0] \supset \mathfrak{F}$$

where q is the G.R. of the formula $\mathfrak{F}$.

Let b be the G.R. of the formula $\mathfrak{A}$.

$$\sim(\exists y_0)(\exists x_0)\{Proof_{P_{\underline{B}}}[x_0, y_0] . Sb[z_0, z_0, y_0]\} \qquad (\mathfrak{A})$$

$Sb[x_0, y_0, z_0]$ is a particular recursion formula such that $Sb[f^{(l)}0, f^{(m)}0, f^{(n)}0]$ holds if and only if n is the G.R. of the result of substituting $f^{(m)}0$ for z_0 in the formula whose G.R. is l at all points where z_0 is free. Let p be the G.R. of the formula $\mathfrak{L}$.

$$\sim(\exists y_0)(\exists x_0)\{Proof_{P_{\underline{B}}}[x_0, y_0] . Sb[f^{(b)}0, f^{(b)}0, y_0]\} \qquad (\mathfrak{L})$$

Then we have as an axiom in $P^{\underline{B}}$

$$(\exists x_0)\, \mathrm{Proof}_{P^{\underline{B}}}[x_0, f^{(p)}0] \supset \mathcal{L}$$

and we can prove in P

$$(x_0)\, Sb\,[f^{(b)}0, f^{(b)}0, x_0] \supset x_0 = f^{(p)}0 \qquad (9.4$$

since $\mathcal{L}$ is the result of substituting $f^{(b)}0$ for Z_0 in $\mathfrak{A}$; whence

$$\sim (\exists y_0)\, \mathrm{Proof}_{P^{\underline{B}}}[y_0, f^{(p)}0] \qquad (9.5)$$

is provable in P. Using (9.4) again we see that $\mathcal{L}$ can be proved in $P^{\underline{B}}$. But if we can prove $\mathcal{L}$ in $P^{\underline{B}}$ then we can prove its provability in $P^{\underline{B}}$, the proof being in P ;i.e. we can prove

$$(\exists x_0)\, \mathrm{Proof}_{P^{\underline{B}}}[x_0, f^{(p)}0]$$

in P (since p is the G.R. of $\mathcal{L}$). But this contradicts (9.5), so that if $\sim \mathfrak{B}[f^{(a-1)}0]$ is true we can prove a contradiction in $P^{\underline{B}}$ or in $P^{\underline{A}*}$. Now I assert that the whole argument up to this point can be carried through formally in the system P, in fact that if c be the G.R. of $\sim(0=0)$ then

$$\sim (a_0)\mathfrak{B}[a_0] \supset (\exists v_0)\, \mathrm{Proof}_{P^{\underline{A}*}}[v_0, f^{(c)}0] \qquad (9.6)$$

is provable in P. I will not attempt to give any more detailed proof of this assertion.

The formula

$$(\exists x_0)\, \mathrm{Proof}_{P\,\mathfrak{A}^*}[x_0, f^{(c)}0] \supset \sim(0=0) \qquad (9.7)$$

is an axiom in $P_{\underline{\mathfrak{A}}}$. Combining (9.6),(9.7) we obtain $(x_0)\,\mathfrak{G}[x_0]$ in $P_{\underline{\mathfrak{A}}}$.

This completeness theorem as usual is of no value. Although it shows for instance that it is possible to prove Fermat's last theorem with Λ_P (if it is true) yet the truth of the theorem would really be assumed by taking a certain formula as an ordinal formula.

That Λ_P is not invariant may be proved easily by our general theorem; alternatively if follows from the fact that in proving our partial completeness theorem we never used ordinals higher than $\omega + 1$. This fact can also be used to prove that Λ_P is not C-K invariant up to $\omega + 2$.

10. The continuum hypothesis. A digression

The methods of § 9 may be applied to problems which are constructive analogues of the continuum hypothesis problem. The continuum hypothesis asserts that $2^{\aleph_0} = \aleph_1$, in other words that if ω_1 is the smallest ordinal α greater than ω such that a series with order type α cannot be put into one-one correspondence with the positive integers, then the ordinals less than ω_1 can be put into one-one correspondence with the subsets of the positive integers. To obtain a constructive analogue of this proposition we may replace the ordinals less than ω_1 either by the ordinal formulae, or by the ordinals represented by them; we may replace the subsets of the positive integers either by the computable sequences of figures 0, 1 or by the description numbers of the machines which compute these sequences. In the manner in which the correspondence is to be set up there is also more than one possibility. Thus even when we use only one kind of ordinal formula there is still great ambiguity as to what the constructive analogue of the continuum hypothesis should be. I shall prove a single result in this connection[23]. A number of others may be proved in the same way.

[23] A suggestion to consider this problem came to me indirectly from F. Bernstein. A related problem was suggested by P. Bernays.

We ask 'Is it possible to find a computable function of ordinal formulae determining a one-one correspondence between the ordinals represented by ordinal formulae and the computable sequences of figures 0, 1?'. More accurately 'Is there a formula $\underline{F}$ such that if $\underline{\Omega}$ is an ordinal formula and n a positive integer then $\underline{F}(\underline{\Omega}, \underline{n})$

is convertible to 1 or to 2, and such that $\underline{F}(\underline{\Omega}, \underline{n})$ conv $\underline{F}(\underline{\Omega}', \underline{n})$, for each positive integer n, if and only if $\underline{\Omega}$ and $\underline{\Omega}'$ represent the same ordinal?'. The answer is 'No', as will be seen to follow from this: there is no formula $\underline{F}$ such that $\underline{F}(\underline{\Omega})$ enumerates a certain sequence of integers (each being 1 or 2) when $\underline{\Omega}$ represents ω and enumerates another sequence when $\underline{\Omega}$ represents 0. If there is such an $\underline{F}$ then there is an a such that $\underline{F}(\underline{\Omega}, \underline{a})$ conv $\underline{F}(\mathrm{Dt}, \underline{a})$ if $\underline{\Omega}$ represents ω but $\underline{F}(\underline{\Omega}, \underline{a})$ and $\underline{F}(\mathrm{Dt}, \underline{a})$ are convertible to different integers (1 or 2) if $\underline{\Omega}$ represents 0. To obtain a contradiction from this we introduce a W.F.F. Gm not unlike Mg. If the machine $\mathcal{M}$ whose D.N. is n has printed 0 by the time the m th complete configuration is reached then $\mathrm{Gm}(\underline{n}, \underline{m})$ conv $\lambda mn.\, m(n, I, 4)$ otherwise $\mathrm{Gm}(\underline{n}, \underline{m})$ conv $\lambda pq.\, \mathrm{Al}(4(P, 2p+2q), 34)$. Now consider $\underline{F}(\mathrm{Dt}, \underline{a})$ and $\underline{F}(\mathrm{Lim}(\mathrm{Gm}(\underline{n})), \underline{a})$. If $\mathcal{M}$ never prints 0 $\mathrm{Lim}(\mathrm{Gm}(\underline{n}))$ represents the ordinal ω. Otherwise it represents 0. Consequently these two formulae are convertible to one another if and only if $\mathcal{M}$ never prints 0. This gives us a means of telling of any machine whether it ever prints 0, which is impossible.

Results of this kind have of course no real relevance for the classical continuum hypothesis.

11. The purpose of ordinal logics.

Mathematical reasoning may be regarded rather schematically as the exercise of a combination of two faculties[24], which we may call

---

[24] We are leaving out of account that most important faculty which distinguishes topics of interest from others; in fact we are regarding the function of the mathematician as simply to determine the truth of falsity of propositions.

---

which we may call intuition and ingenuity. The activity of the intuition consists in making spontaneous judgments which are not the result of conscious trains of reasoning. These judgments are often, but by no means invariably correct (leaving aside the question as to what is meant by 'correct'). Often it is possible to find some other way of verifying the correctness of an intuitive judgment. One may for instance judge that all positive integers are uniquely factorizable into primes; a detailed mathematical argument leads to the same result. It will also involve intuitive judgments, but they will be ones less open to criticism than the original judgment about factorization. I shall not attempt to explain this idea of 'intuition' any more explicitly.

The exercise of ingenuity in mathematics consists in aiding the intuition through suitable arrangements of propositions, and perhaps geometical figures or drawings. It is intended that when these are really well arranged validity of the intuitive steps which are required cannot seriously be doubted.

The parts played by these two faculties differ of course from occasion to occasion, and from mathematician to mathematician. This arbitrariness can be removed by the introduction of a formal logic. The necessity for using the intuition is then greatly reduced by setting down formal rules for carrying out inferences which are always intuitively valid. When working with a formal logic the idea of ingenuity takes a more definite shape. In general a formal logic will be framed so as to admit a considerable variety of possible steps in any stage in a proof. Ingenuity will then determine which steps are the more profitable for the purpose of proving a particular proposition. In pre-Gödel times it was thought by some that it would probably be possible to carry this program to such a point that all the intuitive judgments of mathematics could be replaced by a finite number of these rules. The necessity for intuition would then be entirely eliminated.

In our discussions, however, we have gone to the opposite extreme and eliminated not intuition but ingenuity, and this in spite of the fact that our aim has been in much the same direction. We have been trying to see how far it is possible to eliminate intuition, and leave only ingenuity. We do not mind how much ingenuity is required, and therefore assume it to be available in unlimited supply. In our metamathematical discussions we actually express this assumption rather differently. We are always able to obtain from the rules of a formal logic a method for enumerating the propositions proved by its means. We then imagine that all proofs take the form of a search through this enumeration for the theorem for which a proof is desired. In this way ingenuity is replaced by patience. In these heuristic discussions, however, it is better not to make this reduction.

Owing to the impossibility of finding a formal logic which will wholly eliminate the necessity of using intuition we naturally turn to 'non-constructive' systems of logic with which not all the steps in a proof are mechanical, some being intuitive. An example of a non-constructive logic is afforded by any ordinal logic. When we have an ordinal logic we are in a position to prove number theoretic theorems by the intuitive steps of recognizing formulae as ordinal formulae, and the mechanical steps of carrying out conversions. What properties do we desire a non-constructive logic to have if we are to make use of it for the expression of mathematical proofs? We want it to be quite clear when a step makes use of intuition, and when it is purely formal. The strain put on the intuition should be a minimum. Most important of all, it must be beyond all reasonable doubt that the logic leads to correct results whenever the intuitive steps are correct[25]. It is also desirable that the logic be adequate for the expression of number theoretic theorems, in order that it may be used in metamathematical discussions (cf § 5).

[25] This requirement is very vague. It is not of course intended that the criterion of the correctness of the intuitive steps be the correctness of the final result. The meaning becomes clearer if each intuitive step be regarded as a judgment that a particular proposition is true. In the case of an ordinal logic it is always a judgment that a formula is an ordinal formula, and this is equivalent to judging that a number theoretic proposition is true. In this case then the requirement is that the reputed ordinal logic be an ordinal logic.

Of the particular ordinal logics we have discussed Λ_P and Λ_H certainly will not satisfy us. In the case of Λ_H we are in no better position than with a constructive logic. In the case of Λ_P

(and for that matter also Λ_H) we are by no means certain that we shall never obtain any but true results, because we do not know whether all the number theoretic theorems provable in the system P are true. To take Λ_P as a fundamental non-constructive logic for metamathematical arguments would be most unsound. There remains the system of Church which is free of these objections. It is probably complete (although this would not necessarily mean much) and it is beyond reasonable doubt that it always leads to correct results[26].

[26] This ordinal logic arises from a certain system C_0 in essentially the same way as Λ_P arose from P. By an argument similar to one occurring in § 8 we can show that the ordinal logic leads to correct results if and only if C_0 is valid; the validity of C_0 is proved in Church[1], making use of the results of Church and Rosser [1].

In the next section I propose to describe another ordinal logic, of a very different type, which is suggested by the work of Gentzen, and which should also be adequate for the formalization of number theoretic theorems. In particular it should be suitable for proofs of metamathematical theorems (cf § 5).

12. Gentzen type ordinal logics.

In proving the consistency of a certain system of formal logic Gentzen (Gentzen [1]) has made use of the principle of transfinite induction for ordinals less than ε_0, and suggested that it is to be expected that transfinite induction carried sufficiently far would suffice to solve all problems of consistency. Another suggestion to base systems of logic on transfinite induction has been made by Zermelo (Zermelo [1]). In this section I propose to show how this method of proof may be put into the form of a formal (non-constructive) logic, and afterwards to obtain from it an ordinal logic.

We could express the Gentzen method of proof formally in this way. Let us take the system P and adjoin to it an axiom $\mathfrak{A}_{\underline{\Omega}}$ with the intuitive meaning that the W.F.F. $\underline{\Omega}$ is an ordinal formula, whenever we feel certain that $\underline{\Omega}$ is an ordinal formula. This is a non-constructive system of logic which may easily be put into the form of an ordinal logic. By the method of § 6 we make correspond to the system of logic consisting of P with the axiom $\mathfrak{A}_{\underline{\Omega}}$ adjoined a logic formula $L_{\underline{\Omega}}$: $L_{\underline{\Omega}}$ is an effectively calculable function of $\underline{\Omega}$, and there is therefore a formula Λ_G^1 such that $\Lambda_G^1(\underline{\Omega})$ conv $\underline{\Omega}$ for each formula $\underline{\Omega}$. Λ_G^1 is certainly not an ordinal logic unless P is valid, and therefore consistent. This formalization of Gentzen's idea would therefore not be applicable for the problem with which Gentzen himself was concerned, for he was proving the consistency of a system weaker then P. However, there are other ways in which the Gentzen method of proof can be formalized. I shall explain one, beginning by describing a certain system of symbolic logic.

The symbols of the calculus are f , x , $'$, $_{|}$, 0 , S , R , Γ , Δ , E , $|$, $\odot$, $!$, $($, $)$, $\stackrel{|}{=}$, and the comma ','. We use capital German letters to stand for variable or undetermined sequences of these symbols.

It is to be understood that the relations that we are about to define hold only when compelled to do so by the conditions we lay down. The conditions should be taken together as a simultaneous inductive definition of all the relations involved.

Suffixes

$_{|}$ is a suffix. If $\mathfrak{r}$ is a suffix then $\mathfrak{r}_{|}$ is a suffix.

Indices

$'$ is an index. If $\mathfrak{I}$ is an index then $\mathfrak{I}'$ is an index.

Numerical variables

If $\mathfrak{r}$ is a suffix then $x\mathfrak{r}$ is a numerical variable.

Functional variables

If $\mathfrak{r}$ is a suffix and $\mathfrak{I}$ is an index then $f\mathfrak{r}\mathfrak{I}$ is a functional variable of index $\mathfrak{I}$.

Arguments

$(,)$ is an argument of index $'$. If $(\mathfrak{A})$ is an argument of index $\mathfrak{I}$ and $\mathfrak{T}$ is a term then $(\mathfrak{A}\mathfrak{T},)$ is an argument of index $\mathfrak{I}'$.

Numerals

0 is a numeral.

If $\mathfrak{n}$ is a numeral then $S(,\mathfrak{n},)$ is a numeral.

In metamathematical statements we shall denote the numeral in which S occurs r times by $S^{(r)}(,0,)$.

Expressions of given index

A functional variable of index $\mathfrak{I}$ is an expression of index $\mathfrak{I}$.

R, S are expressions of index $'''$, $''$ respectively.

If $\mathfrak{n}$ is a numeral then it is also an expression of index $'$.

Suppose $\mathfrak{y}$ is an expression of index $\mathfrak{I}$, $\mathfrak{h}$ one of index $\mathfrak{I}'$ and $\mathfrak{R}$ one of index $\mathfrak{I}'''$; then $(\Gamma\mathfrak{y})$ and $(\Delta\mathfrak{y})$ are expressions of index $\mathfrak{I}$, whilst $(E\mathfrak{y})$ and $(\mathfrak{y}\odot\mathfrak{R})$ and $(\mathfrak{y}/\mathfrak{h})$ and $(\mathfrak{y}!\mathfrak{h}!\mathfrak{R})$ are expressions of index $\mathfrak{I}'$.

Function constants

An expression of index $\mathfrak{I}$ in which no functional variable occurs is a function constant of index $\mathfrak{I}$. If in addition R do not occur the expression is called a primitive function constant.

Terms

0 is a term.

Every numerical variable is a term.

If $\mathfrak{y}$ is an expression of index $\mathfrak{I}$ and $(\mathfrak{A})$ is an argument of index $\mathfrak{I}$ then $\mathfrak{F}(\mathfrak{A})$ is a term.

Equations

If $\mathfrak{T}_1$ and $\mathfrak{T}_2$ are terms then $\mathfrak{T}_1 = \mathfrak{T}_2$ is an equation.

Provable equations

We define what is meant by the provable equations relative to a given set of equations as axioms.

(a) The provable equations include all the axioms. The axioms are of the form of equations in which the symbols Γ , Δ , E , $|$, 0 , $!$ do not appear.

(b) If $\mathfrak{F}$ is an expression of index $\mathfrak{I}''$ and $(\mathfrak{A})$ is an argument of index $\mathfrak{I}$ then

$$(\Gamma\mathfrak{F})(\mathfrak{A}\, x_{1}, x_{11},) = \mathfrak{F}(\mathfrak{A}\, x_{11}, x_{1},)$$

is a provable equation.

(c) If $\mathfrak{F}$ is an expression of index $\mathfrak{I}'$, and $(\mathfrak{A})$ is an argument of index $\mathfrak{I}$, then

$$(\Delta\mathfrak{F})(\mathfrak{A}\, x_{1},) = \mathfrak{F}(, x_{1}\, \mathfrak{A})$$

is a provable equation.

(d) If $\mathfrak{F}$ is an expression of index $\mathfrak{I}$, and $(\mathfrak{A})$ is an argument of index $\mathfrak{I}$, then

$$(E\mathfrak{F})(\mathfrak{A}\, x_{1},) = \mathfrak{F}(\mathfrak{A})$$

is a provable equation.

(e) If $\mathfrak{F}$ is an expression of index $\mathfrak{I}$ and $\mathfrak{G}$ is one of index $\mathfrak{I}'$, and $(\mathfrak{A})$ is an argument of index $\mathfrak{I}$, then

$$(\mathfrak{F} \,|\, \mathfrak{G})(\mathfrak{A}) = \mathfrak{G}(\mathfrak{A}\; \mathfrak{F}(\mathfrak{A}),)$$

is a provable equation.

(f) If $\mathfrak{M}$ is an expression of index $'$ then $\mathfrak{M}(,) = \mathfrak{M}$

is a provable equation.

(g) If $\mathfrak{F}$ is an expression of index $\mathfrak{I}$ and $\mathfrak{R}$ one of index $\mathfrak{I}'''$, and $(\mathfrak{A})$ an argument of index $\mathfrak{I}'$, then

$$(\mathfrak{F} \odot \mathfrak{R})(\mathfrak{A}\; 0,) = \mathfrak{F}(\mathfrak{A})$$

and

$$(\mathfrak{F} \odot \mathfrak{R})(\mathfrak{A}\; S(,x_{1},),) = \mathfrak{R}(\mathfrak{A}\; x_{1}, S(,x_{1},), (\mathfrak{F} \odot \mathfrak{R})(\mathfrak{A}\; x_{1},),$$

are provable equations. If in addition $\mathfrak{G}$ is an expression of index $\mathfrak{I}'$ and

$$R(,\mathfrak{G}(\mathfrak{A}\; S(,x_{1},),), x_{1},) = 0$$

is provable then

$$(\mathfrak{F} ! \mathfrak{R} ! \mathfrak{G})(\mathfrak{A}\; S(,x_{1},),) = \mathfrak{R}(\mathfrak{A}\; \mathfrak{G}(\mathfrak{A}\; S(,x_{1},),), S(,x_{1},),$$
$$(\mathfrak{F} ! \mathfrak{R} ! \mathfrak{G})(\mathfrak{A}\; \mathfrak{G}(\mathfrak{A}\; S(,x_{1},),),),)$$

and

$$(\mathfrak{G}\,!\,\mathfrak{R}\,!\,\mathfrak{L})(\mathfrak{A}\;0,) = \mathfrak{G}(\mathfrak{A})$$

are provable.

(h) If $\mathfrak{T}_1 = \mathfrak{T}_2$ and $\mathfrak{T}_3 = \mathfrak{T}_4$ are provable where $\mathfrak{T}_1$, $\mathfrak{T}_2$, $\mathfrak{T}_3$, and $\mathfrak{T}_4$ are terms then $\mathfrak{T}_4 = \mathfrak{T}_3$ and the result of substituting $\mathfrak{T}_3$ for $\mathfrak{T}_4$ at any particular occurrence in $\mathfrak{T}_1 = \mathfrak{T}_2$ are provable equations.

(i) If $\mathfrak{T}_1 = \mathfrak{T}_2$ is a provable equation then the result of substituting any term for a particular numerical variable throughout this equation is provable.

(j) Suppose that $\mathfrak{G}$, $\mathfrak{G}_1$ are expressions of index $\mathfrak{J}'$, that $(\mathfrak{A})$ is an argument of index $\mathfrak{J}$ not containing the numerical variable $\mathfrak{x}$ and that $\mathfrak{G}(\mathfrak{A}\;0,) = \mathfrak{G}_1(\mathfrak{A}\;0,)$ is provable. Also suppose that if we add $\mathfrak{G}(\mathfrak{A}\;\mathfrak{x},) = \mathfrak{G}_1(\mathfrak{A}\;\mathfrak{x},)$ to the axioms and restrict (i) so that it can never be applied to the numerical variable $\mathfrak{x}$ then

$$\mathfrak{G}(\mathfrak{A}\;S(,\mathfrak{x},),) = \mathfrak{G}_1(\mathfrak{A}\;S(,\mathfrak{x},),)$$

becomes a provable equation; in the hypothetical proof of this equation this rule (j) itself may be used provided that a different variable is chosen to take the part of $\mathfrak{x}$.

Under these conditions $\mathfrak{G}(\mathfrak{A}\;\mathfrak{x},) = \mathfrak{G}_1(\mathfrak{A}\;\mathfrak{x},)$ is a provable equation.

(k) Suppose that $\mathfrak{G}$, $\mathfrak{G}_1$, $\mathfrak{L}$ are expressions of index $\mathfrak{J}'$, that $(\mathfrak{A})$ is an argument of index $\mathfrak{J}$ not containing the numerical variable $\mathfrak{x}$ and that $\mathfrak{G}(\mathfrak{r}\,0,) = \mathfrak{G}_1(\mathfrak{r}\,0,)$ and $R(,\mathfrak{H}(\mathfrak{r}\,S(,\mathfrak{x},),),S(,\mathfrak{x},),) = 0$ are provable equations.

Suppose also that if we add

$$\mathfrak{G}(\mathfrak{r}\,\mathfrak{H}(\mathfrak{r}\,S(,\mathfrak{x},),),) = \mathfrak{G}_1(\mathfrak{r}\,\mathfrak{H}(\mathfrak{r}\,S(,\mathfrak{x},),),)$$

to the axioms, and again restrict (i) so as not to apply to $\mathfrak{x}$ then

$$\mathfrak{G}(\mathfrak{r}\,\mathfrak{x},) = \mathfrak{G}_1(\mathfrak{r}\,\mathfrak{x},) \qquad (12.1)$$

becomes a provable equation; in the hypothetical proof of (12.1) the rule (k) may be used if a different variable takes the part of $\mathfrak{x}$.

Under these conditions (12.1) is a provable equation.

We have now completed the definition of a provable equation relative to a given set of axioms. Next we shall show how to obtain an ordinal logic from this calculus. The first step is to set up a correspondence between some of the equations and number theoretic theorems, in other words to show how they can be interpreted as number theoretic theorems. Let $\mathfrak{G}$ be a primitive function constant of index III. $\mathfrak{G}$ describes a certain primitive recursive function $\varphi(m,n)$, determined by the condition that for all m, n the equation

$$\mathfrak{G}(,S^{(m)}(,0,),S^{(n)}(,0,),) = S^{(\varphi(m,n))}(,0,)$$

shall be provable without using the axioms (a). Suppose also that $\mathfrak{H}$ is an expression of index I. Then to the equation

$$\mathfrak{G}(,x_1,\mathfrak{H}(x_1,),) = 0$$

we make correspond the number theoretic theorem which asserts that for each natural number m there is a natural number n such that $\varphi(m,n) = 0$. (The circumstances that there is more than one equation to represent each number theoretic theorem could be avoided by a trivial modification of the calculus.)

Now let us suppose some definite method is chosen for describing the sets of axioms by means of positive integers, the null set of axioms being described by the integer 1. By an argument used in § 6 there is a W.F.F. Σ such that if r is the integer describing a set A of axioms then $\Sigma(\underline{r})$ is a logic formula enabling us to prove just those number theoretic theorems which are associated with equations provable with the above described calculus, the axioms being just those described by the number r.

I shall show two ways in which the construction of the ordinal logic may be completed.

In the first method we make use of the theory of general recursive functions (Kleene [2]). Let us consider all equations of the form

$$R(,S^{(m)}(,0,),S^{(n)}(,0,),)=S^{(p)}(,0,) \qquad (12.2)$$

which are obtainable from the axioms by the use of rules (h), (i). It is a consequence of the theorem of equivalence of λ-definable and general recursive function (Kleene [5]) that if $r(m,n)$ is any λ-definable function of two variables then we can choose the axioms so that (12.2) with $p = r(m,n)$ is obtainable in this way for each pair of natural numbers m, n, and no equation of the form

$$S^{(m)}(,0,) = S^{(n)}(,0,) \quad (m \neq n) \qquad (12.3)$$

is obtainable. In particular this is the case if $r(m,n)$ is defined by the condition that

$$\underline{\Omega}(\underline{m},\underline{n}) \text{ conv } S(\underline{p}) \text{ implies } p = r(m,n)$$
$$r(0,n) = 0 \text{ all } n > 0, \quad r(0,0) = 2$$

where $\underline{\Omega}$ is an ordinal formula. There is a method for obtaining the axioms given the ordinal formula, and consequently a formula Rec such that for any ordinal formula $\underline{\Omega}$, $Rec(\underline{\Omega})$ conv $\underline{m}$ where m is the integer describing the set of axioms corresponding to $\underline{\Omega}$. Then the formula

$$\Lambda_G^2 \rightarrow \lambda\omega.\ \Sigma(Rec(\omega))$$

is an ordinal logic. Let us leave the proof of this aside for the present.

Our second ordinal logic is to be constructed by a method not unlike the one we used in constructing Λ_P. We begin by assigning ordinal formulae to all sets of axioms satisfying certain conditions. For this purpose we again consider that part of the calculus which is obtained by restricting 'expressions' to be functional variables or R or S and restricting the meaning of 'term' accordingly; the new provable equations are given by conditions (a), (h), (i), together with an extra condition (l)

(1) The equation

$$R(, 0, S(, x_1,),) = 0$$

is provable.

We could design a machine which would obtain all equations of the form (12.2), with $m \neq n$, provable in this sense, and all of the form (12.3), except that it would cease to obtain any more equations when it had once obtained one of the latter 'contradictory' equations. From the description of the machine we obtain a formula $\underline{\Omega}$ such that

$\underline{\Omega}(\underline{m}, \underline{n})$ conv 2 if $R(, S^{(m-1)}(, 0,), S^{(n-1)}(, 0,),) = 0$ is obtained by the machine

$\underline{\Omega}(\underline{m}, \underline{n})$ conv 1 if $R(, S^{(n-1)}(, 0,), S^{(m-1)}(, 0,),) = 0$ is obtained by the machine

$\underline{\Omega}(\underline{m}, \underline{m})$ conv 3 always.

The formula $\underline{\Omega}$ is an effectively calculable function of the set of axioms, and therefore also of w; consequently there is a formula M such that $M(\underline{w})$ conv $\underline{\Omega}$ when w describes the set of axioms. Now let Cm be a formula such that if b is the G.R. of a formula $M(\underline{w})$ then $Cm(\underline{b})$ conv $\underline{w}$, but otherwise $Cm(\underline{b})$ conv 1. Let

$$\Lambda_G^3 \rightarrow \lambda \varpi a.\, \Gamma(\lambda u.\, \Sigma(Cm(Th(\varpi, u)), a)$$

Then $\Lambda_G^3(\underline{\Omega}, \underline{A})$ conv 2 if and only if $\underline{\Omega}$ conv $M(\underline{w})$ where w describes a set of axioms which, taken with our calculus, suffices to prove the equation which is, roughly speaking, equivalent to '$\underline{A}$ is dual'. To prove that Λ_G^3 is an ordinal logic it suffices to prove that the calculus with the axioms described by w proves only true number theoretic theorems when $\underline{\Omega}$ is an ordinal formula. This condition on w may also be expressed in this way. Let us put $m \ll n$ if we can prove $R(, S^{(m)}(, 0,), S^{(n)}(, 0,),) = 0$ with (a), (h), (i), (l): the condition is that $m \ll n$ be a well ordering of the natural numbers and that no contradictory equation (12.3) be provable with the same rules (a), (h), (i), (l). Let us say that such a set of axioms is <u>admissible</u>. Λ_G^3 is an ordinal logic if the calculus leads to none but true number theoretic theorems when an admissible set of axioms is used.

In the case of Λ_G^2, $Rc(\underline{\Omega})$ describes an admissible set of axioms whenever $\underline{\Omega}$ is an ordinal formula. Λ_G^2 will therefore be an ordinal logic if the calculus leads to correct results when admissible axioms are used.

To prove that admissible axioms have this property I shall not attempt to do more than show how interpretations can be given to the equations of the calculus so that the rules of inference (a) - (k) become intuitively valid methods of deduction, and so that the interpretation agrees with our convention regarding number theoretic theorems.

Each expression is the name of a function, which may be only partially defined. The expression S corresponds simply to the successor function. If $\mathfrak{G}$ is either R or a functional variable and is of index $\mathfrak{J}$ ($p+1$ symbols in the index) then it corresponds to a function g of p natural numbers defined as follows. If

$$\mathfrak{J}(,S^{(r_1)}(,0,),S^{(r_2)}(,0,),\ldots,S^{(r_p)}(,0,),)=S^{(l)}(,0,)$$

is provable by the use of (a), (h), (i), (l) only, then $g(r_1,\ldots,r_p)$ has the value l. It may not be defined for all arguments, but its value is always unique, for otherwise we could prove a 'contradictory' equation and $M(\underline{m})$ would then not be an ordinal formula. The functions corresponding to the other expressions are essentially defined by (b) - (f). For example if g is the function corresponding to $\mathfrak{J}$ and g' that corresponding to $(\Gamma\,\mathfrak{J})$ then

$$g'(r_1,r_2,\ldots,r_p,l,m)=g(r_1,r_2,\ldots,r_p,m,l)$$

The values of the functions are clearly unique (when defined at all) if given by one of (b) - (e). The case (f) is less obvious since the function defined appears also in the definiens. I shall not treat the case of $(\mathfrak{J}\odot\mathfrak{H})$ as this is the well known definition by primitive recursion, but let us show the values of the function corresponding to $(\mathfrak{J}\,!\,\mathfrak{R}\,!\,\mathfrak{H})$ are unique. Without loss of generality we may suppose that $(\mathfrak{R})$ is of index $'$. We have then to show that if $h(m)$ is the function corresponding to $\mathfrak{H}$ and $r(m,n)$ that corresponding to $\mathfrak{R}$, and $k(u,v,w)$ a given function and a a given natural number then the equations

$$l(0)=a \qquad \alpha)$$

$$l(m+1)=k(h(m+1),m+1,l(h(m+1))) \qquad \beta)$$

do not ever assign two different values for the function $l(m)$. Consider those values of r for which we obtain more than one value of $l(r)$, and suppose that there is at least one such. Clearly 0 is not one for $l(0)$ can only be defined by $\alpha)$. As the relation $<<$ is a well ordering there is an integer r_0 such that $r_0 > 0$, $l(r_0)$ is not unique, and if $s \neq r_0$ and $l(s)$ is not unique then $r_0 << s$. Putting $s = h(r_0)$ we find also $s << r_0$ which is impossible. There is therefore no value for which we obtain more than one value for the function $l(r)$.

Our interpretation of expressions as functions give us an immediate interpretation for equations with no numerical variables. In general we interpret an equation with numerical variables as the conjunction of all equations obtainable by replacing the variables by numerals. With this interpretation (h), (i) are seen to be valid methods of proof. In (j) the provability of

$$\mathfrak{G}(\mathfrak{A}\, S(,x_{,},),) = \mathfrak{G}_1(\mathfrak{A}\, S(,x_{,},),)$$

when $\mathfrak{G}(\mathfrak{A}\, x_{,},) = \mathfrak{G}_1(\mathfrak{A}\, x_{,},)$ is assumed to be interpreted as meaning that the implication between these equations holds for all substitutions of numerals for $x_{,}$. To justify this one should satisfy oneself that these implications always hold when the hypothetical proof can be carried out. The rule of procedure (j) is now seen to be simply mathematical induction. The rule (k) is a form of transfinite induction. In proving the validity of (k) we may again suppose $(\mathfrak{A})$ is of index $'$. Let $r(m,n)$, $g(m)$, $g_1(m)$, $h(n)$ be the functions corresponding respectively to $\mathfrak{R}$, $\mathfrak{G}$, $\mathfrak{G}_1$, $\mathfrak{H}$.

We shall prove that if $g(0)=g_1(0)$ and $r(h(n),n)=0$ for each positive integer n and $g(n+1)=g_1(n+1)$ whenever $g(h(n+1))=g_1(h(n+1))$ then $g(n)=g_1(n)$ for each natural number n. We consider the class of integers n for which $g(n)=g_1(n)$ is not true. If the class is not void it has a positive member n_0 which precedes all other members in the well ordering $\ll$. But $h(n_0)$ is another member of the class, for otherwise we should have $g(h(n_0))=g_1(h(n_0))$ and therefore $g(n_0)=g_1(n_0)$ i.e. n_0 would not be in the class. This implies $n_0 \ll h(n_0)$ contrary to $r(h(n_0),n_0)=0$. The class is therefore void.

It should be noticed that we do not really need to make use of the fact that $\underline{\Omega}$ is an ordinal formula. It suffices that $\underline{\Omega}$ should satisfy conditions (a) - (e) (p.29) for ordinal formulae, and in place of (f) satisfy (f').

(f') There is no formula $\underline{I}$ such that $\underline{I}(\underline{n})$ is convertible to a formula representing a positive integer for each positive integer n, and such that $\underline{\Omega}(\underline{I}(\underline{n}),\underline{n})$ conv 2, for each positive integer n for which $\underline{\Omega}(\underline{n},\underline{n})$ conv 3.

The problem as to whether a formula satisfies conditions (a) - (e), (f') is number theoretic. If we use formulae satisfying these conditions instead of ordinal formulae with Λ_G^3 we have a non-constructive logic with certain advantages over ordinal logics. The intuitive judgments that must be made are all judgments of the truth of number theoretic theorems. We have seen in § 9 that the connection of ordinal logics with the classical theory of ordinals is quite superficial. There seem to be good reasons therefore for giving attention to ordinal formulae in this modified sense.

The ordinal logic Λ_G^3 appears to be adequate for most purposes. It should for instance be possible to carry out Gentzen's proof of consistency of number theory, or the proof of the uniqueness of the normal form of a well-formed formula (Church and Rosser [1]) with our calculus and a fairly simple set of axioms. How far this is the case can of course only be determined by experiment.

One would prefer that a non-constructive system of logic based on transfinite induction were rather simpler than the one we have described. In particular it would seem that it should be possible to eliminate the necessity of stating explicitly the validity of definitions by primitive recursions, as this principle itself can been shown to be valid by transfinite induction. It is possible to make such modifications in the system, even in such a way that the resulting system is still complete, but no real advantage is gained by doing so. The effect is always, so far as I know, to restrict the class of formulae provable with a given set of axioms, so that we obtain no theorems but trivial restatements of the axioms. We have therefore to compromise between simplicity and comprehensiveness.

Index of definitions

No attempt is being made to list underlined formulae as their meanings are not always constant throughout the paper. Abbreviations for definite well-formed formulae are listed alphabetically.

Miscellaneous (in order of appearance)

Bibliography

Alonzo Church

[1]. A proof of freedom from contradiction, Proc. Nat. Acad. Sci., 21 (1935) 275-281.
[2]. Mathematical logic, Lectures at Princeton University (1935-6), mimeographed, 113 pp.
[3]. An unsolvable problem of elementary number theory, Am. Jour. Math., 58 (1936) 345-363.
[4]. The constructive second number class, forthcoming in Bull. Am. Math. Soc.

G. Gentzen

[1]. Die Widerspruchsfreiheit der reinen Zahlentheorie, Math. Annalen, 112 (1936) 493-565.

K. Gödel

[1]. Über formal unentscheidbare Satze der Principia Mathematica und verwandter Systeme, Monatshefte Math. Phys., 38 (1931) 173-198.
[2]. On undecidable propositions of formal mathematical systems, Lectures at the Institute for Advanced Study, Princeton, N. J. 1934, mimeographed, 30 pp.

S. C. Kleene

[1]. A theory of positive integers in formal logic, Am. Jour. Math. 57 (1935) 153-173 and 219-244.
[2]. General recursive functions of natural numbers, Math. Annalen, 112 (1935-6) 727-742.
[3]. λ-definability and recursiveness, Duke Math. Jour., 2 (1936) 340-353.

E. L. Post

[1]. Finite combinatory processes - formulation 1, Jour. Symbolic Logic, 1 (1936) 103-105.

J. B. Rosser

[1]. Gödel theorems for non-constructive logics, Jour. Symbolic Logic, 2 (1937) 129-137.

A. Tarski

[1]. Der Wahrheitsbegriff in den formalisierten Sprachen, Studia

Philosophica, 1 (1936) 261-405, (translation from the original paper in Polish dated 1933).

A. M. Turing

[1]. On computable numbers, with an application to the Entscheidungsproblem, Proc. Lond. Math. Soc. (ser 2) 42 (1936-7) 230-265. A correction to this paper has appeared in the same journal 43 (1937) 544-546.
[2]. Computability and λ-definability, Jour. Symbolic Logic, 2 (1937) 153-163.

E. Zermelo

[1]. Grundlagen einer allgemeiner Theorie der mathematischen Satzsysteme I, Fund. Math., 25 (1935) 136-146.

Alonzo Church and S. C. Kleene

[1] Formal definitions in the theory of ordinal numbers, Fund. Math. 28 (1936) 11-21.

Alonzo Church and J. B. Rosser

[1]. Some properties of conversion, Trans. Am. Math. Soc., 39 (1936) 472-482.

D. Hilbert and W. Ackermann

[1]. Grundzüge der theoretischen Logik, (2nd edn. revised Berlin, 1938) 130 pp.

A. N. Whitehead and Bertrand Russell

[1]. Principia Mathematica, (2nd edition, Cambridge, 1925-1927), 3 vols.

D. Hilbert

[1]. Uber das unendliche, Math. Annalen, 95 (1926), 161-190

A Remarkable Bibliography

The bibliography of Alan Turing's PhD thesis is most remarkable in that every author cited is one of the most influential logicians of all time.

Wilhelm Ackermann, 1896–1962, earned his PhD from Hilbert in 1925, and is known for many results in logic, particularly the fast-growing Ackermann function, which is computable but not primitive-recursive.

Alonzo Church, 1903–1995, created the λ-calculus, on which almost all modern programming languages are based. He is buried in the Princeton Cemetery.

Gerhard Gentzen, 1909–1945, was a pioneer in proof theory. He died of starvation at the end of World War II after being arrested as a German national in Prague.

Kurt Gödel, 1906–1978, stunned the mathematical world with his great incompleteness results of 1931. Shortly thereafter he moved to the Institute for Advanced Study in Princeton; he is buried in the Princeton Cemetery.

David Hilbert, 1862–1943, the most influential mathematician of the early twentieth century, was chairman of mathematics at Göttingen from 1895. He contributed greatly to the increased rigor of mathematics, and in 1920 posed the problems of whether mathematical truth and proof could always be derived by a mechanical procedure. Between 1931 and 1936, Gödel, Church, and Turing demonstrated conclusively that the answer is no.

Stephen Kleene, 1909–1994, earned his PhD at Princeton under Alonzo Church in 1934, and was extremely influential in the creation of modern recursive function theory. He spent most of his career at the University of Wisconsin.

Emil Post, 1897–1954, was a mathematician and logician best known for helping create the field of computability theory. Having emigrated from Poland to New York as a child, he completed a PhD at Columbia University in 1920 and then a postdoc at Princeton. In the late 1940s he was the first to recognize the importance of Turing's two-page digression (pp. 18–19) about oracle machines.

J. Barkley Rosser, 1907–1989, who earned his PhD under Alonzo Church in 1934, is known for the Church-Rosser theorem (confluence of reduction) and other important early results in λ-calculus; later in his career he worked on prime number computations, the Riemann zeta function, and numerical methods, as well as logic.

Bertrand Russell, 1872–1970, was a philosopher, mathematician, logician, and social critic, and a PhD student of Alfred North Whitehead's. Their joint work, *Principia Mathematica* (1910), was an attempt to derive real mathematics in a fully formal, logical way. Practically it was not a great success, but the attempt was enormously influential—the title of Gödel's great 1931 result was "On formally undecidable propositions of *Principia Mathematica* and related systems."

Alfred Tarski, 1901–1983, one of the greatest logicians of the twentieth century, emigrated from Poland in 1939 and taught at the University of California, Berkeley for forty years. "Along with his contemporary, Kurt Gödel, he changed the face of logic in the twentieth century, especially through his work on the concept of truth and the theory of models" (Feferman).

Alan Turing, 1912–1954, an English mathematician, earned his PhD under Alonzo Church in 1938. He is widely considered to be the father of computer science and artificial intelligence.

Alfred North Whitehead, 1861–1947, was an English mathematician, logician, and philosopher.

Ernst Zermelo, 1871–1953, the developer of modern set theory, worked in Berlin, Göttingen, Zurich, and Freiburg, except during the years 1935–1945, when he resigned his position in disapproval of the Nazi regime.

Contributors

ANDREW W. APPEL is Eugene Higgins Professor and Chairman of the Department of Computer Science at Princeton University. His research is in computer security, programming languages and compilers, automated theorem proving, and technology policy. He received his A.B. summa cum laude in physics from Princeton, and his PhD in computer science from Carnegie Mellon University.

SOLOMON FEFERMAN is the Patrick Suppes Family Professor of Humanities and Sciences, Emeritus, and Professor of Mathematics and Philosophy, Emeritus, at Stanford University. His main areas of research are in mathematical logic, the foundations of mathematics, and the history of twentieth-century logic. He is the editor-in-chief of the five-volume edition of Kurt Gödel's *Collected Works* and is the coauthor, with Anita Burdman Feferman, of *Alfred Tarski: Life and Logic*.